MW00573682

Fascinating Life Sciences

This interdisciplinary series brings together the most essential and captivating topics in the life sciences. They range from the plant sciences to zoology, from the microbiome to macrobiome, and from basic biology to biotechnology. The series not only highlights fascinating research; it also discusses major challenges associated with the life sciences and related disciplines and outlines future research directions. Individual volumes provide in-depth information, are richly illustrated with photographs, illustrations, and maps, and feature suggestions for further reading or glossaries where appropriate.

Interested researchers in all areas of the life sciences, as well as biology enthusiasts, will find the series' interdisciplinary focus and highly readable volumes especially appealing.

Claudio Campagna • Daniel Guevara

Speaking of Forms of Life

The Language of Conservation

Claudio Campagna
Language of Conservation Project &
Wildlife Conservation Society
Buenos Aires, Argentina

Daniel Guevara
Language of Conservation Project &
University of California
Santa Cruz, CA, USA

ISSN 2509-6745 ISSN 2509-6753 (electronic)
Fascinating Life Sciences
ISBN 978-3-031-34533-3 ISBN 978-3-031-34534-0 (eBook)
https://doi.org/10.1007/978-3-031-34534-0

Cover illustration by Eugenia Zavattieri: Pufferfish nest (see: https://www.bbc.co.uk/programmes/p029nb9g).

This Springer imprint is published by the registered company Springer Nature Switzerland AG
The registered company address is: Gewerbestrasse 11, 6330 Cham, Switzerland

Paper in this product is recyclable.

"*Speaking of Forms of Life* helps us confront the fact that language often obscures, rather than facilitates, our understanding of living things. Our language reflects past misperceptions, current ignorance, and our shockingly limited intellectual ability to comprehend where we are in space and time, and who we are with on this strange planet with its soap-bubble coating of life. The language in use has abetted and accelerated the catastrophic course we daily continue to choose. *Speaking of Forms of Life* shows us why and how this must change. It is a crucial revelation that we must heed, because our species alone can consider changing course—and our species alone must do so."

—Carl Safina, *MacArthur Fellow and Carl Safina Research Chair for Nature and Humanity, Stony Brook University*

"This is an essential book for anyone who cares about conservation and is concerned about the frightening pace of extinctions. The great theme of the book is the importance of the language in which we think and talk about living beings. Claudio Campagna and Daniel Guevara make available for conservationists the groundbreaking work of Philippa Foot and Michael Thompson on natural goodness and on our thought about living beings. They show how it can be brought to bear on the threats that confront conservation and on disputes that may seem irresolvable. I recommend *Speaking of Forms of Life* as strongly as I can."

—Cora Diamond, *William R. Kenan Jr. Professor Emerita of Philosophy, Professor of Law, and University Professor, University of Virginia*

"Our current conservation language is shot through with economic platitudes that don't explain anyone's real motives for conserving biodiversity. Campagna and Guevara's new book lays out a convincing alternative, grounding conservation goals in the objective goodness of life's many ways of flourishing. *Speaking of Forms of Life* is inspiring, informative, well-grounded in the relevant philosophy and conservation literatures but never pedantic. Both activists and thinkers will find much of value in *Speaking of Forms of Life*."

—Philip Cafaro, *Professor of Philosophy, Colorado State University*

"Since Linnaeus, biologists have been cataloguing all forms of life, assigning names without much consideration how each one came to be what it is and to do what it does. In *Speaking of Forms of Life*, Campagna and Guevara urge us to consider the cost of our propensity to categorize, to assign value to life forms in taxonomic or economic terms. In doing so, they argue, we forget where we started: valuing species for what they are, for their interactions with other species and with their habitats. Adopting language that considers species' "natural goodness," however, provides a much-needed moral framework for modern environmentalism. Everyone who cares about the extinction crisis that we are currently experiencing—about the decisions that we and our children will necessarily have to make—should read this book."

—Beth Shapiro, *Howard Hughes Medical Institute Investigator and Professor of Ecology and Evolutionary Biology, University of California, Santa Cruz*

"Our biosphere is a miracle. Communicating the sense of awe and wonder that it instills in many of us is essential to shift from wanton destruction of our natural world for short-term economic profit to health and prosperity for all creatures—including us humans. *Speaking of Forms of Life* is the first book that unveils how the way we speak to each other affects the way we value and care for our planet. An essential read for everyone who cares about the future of nature and humanity."

—Enric Sala, *National Geographic Explorer in Residence and Hubbard Medalist*

This book is dedicated to the memory of Philippa Foot, and to all those who practice life form conservation, whether they know it or not.

Preface for Conservationists

I grew up as an urban dweller, near Buenos Aires, with the ocean 400 km away, the mountains 1000. A couple of grazing cows in the grasslands of the pampas were enough to call "nature." I was far from a childhood of long walks with binoculars or of excursions among wildflowers. The animals I knew and appreciated were domestic; as for plants, what mattered were those in my father's garden, which made for a good salad.

Nature transformed into a distinct place for me when I began to imagine it through the eyes and words of others. Jacques Cousteau was first and then Konrad Lorenz. At the time, their books and documentaries unleashed a feeling of respectful admiration for all living things. Lorenz inspired the possibility of comprehending animals, a step toward sympathizing with their struggles against human threats. Together, Cousteau's images of subaquatic life and 'Lorenzian'[1] swims among wild geese contributed to the development of aesthetic appreciation for the natural world and to a sense of compassion for it, evoked by the sad relationship that humanity had established with other living things.

Admiration, sympathy, compassion, and aesthetic appreciation are virtues, and some of them, in the right context, are ethical virtues. In the work before you we maintain that the ethical virtues are essential to the practice of conservation. In this respect, our work is not alone [1–7]. Thus, if my early personal history has any relevance here, it is because it suggests, perhaps contrary to academic understanding, that words and images can unlock the sentiments of virtue. A book may work, some pictures too. The adventurous stories and the emotions evoked by the Calypso surrounded by dolphins, even if only witnessed through a black and white television screen, prompted sufficient reason for developing a unique quality of appreciation for *The Living Sea*, the title of the first Cousteau book I read. Years later, the same effect was produced by reading that whales communicated through song, and by listening to a taped recording of them; science could be inspiring [8].

[1] We use quotation marks in the standard way when citing a speaker or text directly, or indicating mention, as opposed to use, of a term. But we use single quotes also as scare quotes to indicate technical, non-standard, or contentious use of a term (usually only upon its first introduction, but occasionally when context might require flagging it again).

In time, I myself became an ethologist and then a marine conservationist. For a long time, I had been disconcerted by the extinction of the passenger pigeon. Today, at least another 9000 species, categorized as Critically Endangered by the International Union for Conservation of Nature (IUCN), are at risk of the same fate [9]. But what is most striking about the passenger pigeon is the surreal fact that its extraordinary abundance did not prevent it from being wiped out, but in fact led to our extinguishing it. It is a great embarrassment to understand how human will is implicated in this. A failure of our most exceptional trait, human reason, contributes to the problem—as though something deeply mistaken drives the drama of humanity.

Not surprisingly, my early heroes were aware of conflict between humans and nature [10], although they did not, I now understand, get to the heart of the matter. Aldo Leopold had glimpsed the essence of the problem in his observation that language hits a limit when it tries to capture certain values in nature [4]. In this book, we investigate this insight in depth, with careful attention to the connection between the critical concepts of language, life, and values. It would have helped my own career to have understood this connection earlier. Instead, for years I used the tools of science to confront the debates and dilemmas of ethicists. It seems that my maturation process is not special, as explained by E.O. Wilson,

> When very little is known about an important subject, the questions people raise are almost inevitably ethical. Then as knowledge grows, they become more concerned with information and amoral, in other words more narrowly intellectual. Finally, as understanding becomes sufficiently complete, the questions turn ethical again. Environmentalism is now passing from the first to the second phase, and there is reason to hope that it will proceed directly on to the third [11, p. 119].

In the essay that follows, the reader will find a slight variation on this process: when so little is known about living things, human beings can only wonder what they do, how they function, and what their form of life is. As this knowledge is perfected, questions of value emerge, including the question of what our relationship should be to the rest of life. The conservation movement today seems to have stalled in the second stage of Wilson's process, of acquiring scientific knowledge. When it comes to facing ethical issues, it makes pronouncements about what is good or bad, but it does so without rigorous grounds or lasting success. Our essay argues that both stages merge into one: the description of living things—of what they have and do to satisfy their primary needs and functions—implicitly and logically contains the standard for understanding the system of values on which to base an ethical relationship to the other life forms. Consider: "Sharks move in the water by using their fins." This simple phrase states a commonly known fact about sharks. Yes, but much more, inasmuch as it also identifies a primary value in the creature's satisfaction of the needs of its form of life. Fact and value are integrated in the same judgment. In what follows we give close attention to the logic of these apparently simple judgments, and to the reasons why the practice of conservation must be based on the language that expresses them: the familiar language of natural history.

Finally, I want to urge my fellow conservationists to be aware of the language, the form of expression, they use to describe life and the living things, and of the reasons they intervene on behalf of nature to redress the harm or wrongs they see there. Language, as we understand and explain in this essay, is crucial to the development and strengthening of the virtues related to our properly valuing living beings, and to stopping the extinctions we provoke.[2] I believe that having adopted the language of natural history, in the image and likeness of Lorenz and Cousteau, led me to affirm and adopt the disposition of character necessary for the protection of living things. I deeply agree with what Philip Cafaro suggested, in an insightful paper [1], where he says that "studying natural history helps make us more virtuous; that is, better and happier people." Many practitioners would agree that their reason for being conservationists is based on what living beings are and do. In the work that follows, we argue that this is indeed the primary reason for the practice of conservation, as grounded in the unique logic of the language used to express what a living being is and does according to its form of life. The use of the naturalistic language alone does not guarantee adopting the necessary decisions to practice conservation for the sake of all the living creatures, but will place us in the right path. Conservationists tend to be practical and to prioritize actions without necessarily digging deeply into the values that underpin them. As a result, countless competing value systems, driven by their own language and logic, only keep us from understanding the significance of the extinctions we provoke.

The expectation is that this book will contribute to the efforts of environmental ethicists, naturalists, conservation biologists, writers, or journalists that we admire (e.g., [11–17]). We all face an unforgiving reality check: more words, more budget, more politics, more science, and more economics would not turn 9000 potential extinctions into 0, or even 8999. Something deep, transformative, has to happen at the level of what Bruno Latour calls "collectives." Standards of good and bad, the acceptable or unacceptable, must change. Values must be replaced. Language is both the disruptive and constructive lever.

There is a concept of "social impetus" that refers to what some argue is a transformative power in human action [18]. I would argue that the impetus for the conservation movement to secure and expand its ethical momentum is to be found in the way we speak about extinctions. Perhaps, some ideas in this book will help 'rescue' those who, after years of practice, or even before starting one, are seeking to align mind and heart, and the authority to declare unconditionally, that they value living beings for what they are and do according to their forms of life, and that this is the primary reason to conserve them and to be a conservationist. Paraphrasing the great Darwin, there is a grandeur in the description of life, with its diverse forms, conceived by an original mechanism of nature, as if with the intention of grounding in these very forms the reason to value and protect them. The language we call "the language of forms of life," which describes life and expresses the primary standard

---

[2] We use "provoke" in an attempt to emphasize the agency or intentionality in the extinctions we cause.

for making value judgments about it, is the language of Darwin and all naturalists, of medical science (where I first learned to master it), and of biology itself—made possible by a unique, objective, logical grammar behind everything we can say or think about living things.

Buenos Aires, Argentina

Claudio Campagna

Preface for Philosophers

In the conservation of nature and, more broadly, environmentalism, we confront philosophical issues for which we hardly have the language to express or explore what seems to concern us. Indeed, the very terms "environment" and "nature" illustrate the point; in the intended senses, they are exceedingly flexible terms, signifying almost anything and everything, including the air, land, and sea, their beauty and usefulness, and all living things, in all their great diversity of forms. Perhaps some, when using these terms, mean to exclude human beings and their artifacts, practices, or institutions. Still, the exclusion has more to do with identifying human beings as a problem for the 'rest of nature (or the environment),' rather than with excluding them from nature. Some might say that the language of 'sustainable development' has given us the needed precision, but as we argue in what follows, this is a tragic mistake. What is done in the name of sustainable development often does the opposite of what we would expect from the conservation of nature or concern for the environment [19].

The broad topics of 'conservation' and 'environmentalism' get a kind of organizing theme when they pose the question of the environment or nature as one about the value of non-human beings, or about our 'relationship' to non-human nature, where this is intended to raise ethical issues. But here too almost anything goes. It is true that the emphasis is commonly on the non-human animals 'closest' to us (e.g., 'sentient' or 'intelligent'), or certain flora we tend to favor (forests, for example), and their value—instrumental, aesthetic, spiritual, and intrinsic, among others. And some are even willing to extend these questions of value and ethical relationship (including talk of applying 'rights') to deserts and oceans, and all that is in them, and more. Of course, many of us are concerned with species extinction and biodiversity, quite generally. Aldo Leopold's celebrated "Land Ethic" is the classic attempt to unite our intelligence with our heart, in our struggles with these issues [4]. But he was aware of the limitations of what he was doing in trying to articulate "values as yet uncaptured by language," as he calls them, in his characteristically prescient and elegant style.

We offer, in the work before you, the key features of a language that we think shows considerable promise for representing the values in nature that have eluded proper philosophical articulation so far, especially with respect to the conservation of species. We believe that attention to the logic and grammar of this language, the language of natural history (mastered already by many conservationists and familiar to all of us from an early age), has the potential to change conservation as we know it, on rigorous philosophical grounds. It offers the kind of radical rethinking and fresh start that we think is necessary after so many false starts and conflicting movements and ideologies in conservation, well-meaning and stimulating as so many of them have been.

Gary Snyder has observed that:

> … no political movement in the United States that came out of left field with so little beginning public support has had as much effect on the whole American political and economic system in such a short time as the environmental movement [20, p. 139].

He said that about 40 years ago and the same can now be said about the movement's effect on the rest of the world. However, I believe that what Snyder called our attention to—when the movement was just coming out of left field and out of the blue—is something we should have given more thought to before getting started on, or continuing with, any environmental philosophizing. The philosophical questions urged by the environmental movement have hardly, if ever, been confronted in the history of philosophy—at least in any Western tradition. Among these are ethical questions about the value of life, in all its forms, or of the relation of human beings to the rest of nature, asked with the intention of challenging traditional, near universal, anthropocentric assumptions about value. And the movement's remarkable success story has masked the challenges posed for philosophical reflection on all this.

The movement Snyder is referring to, and that most of us call "environmentalism" today, had its grassroots origins in the West: the west coast of California and Pacific Northwest (which some would be happy to call the fringe)—John Muir, David Brower, and Gary Snyder, among the prime movers. It quickly became a worldwide movement, far beyond its impact on the USA, with its many national and international societies, NGOs, governmental agencies, and laws. It just as quickly became a staple of academic life, with the proliferation of standard college courses, textbooks, and programs in environmentalism across the letters and sciences, at both undergraduate and graduate levels. But in all the philosophizing about the environment or nature that has transpired since, we have not appreciated the full significance of all this success. Among the best evidence that we have not done so is that environmental philosophizing has almost always tried to apply one or another well-known philosophical tradition to these new questions, or else applied a pluralistic mix of traditions, taking what seemed useful from this one here (Kantian or utilitarian, say) and another there (e.g., Buddhist or Spinozistic). For the movement to have developed in this way is understandable because where else are we to look for guidance, if not to what we already know, or what has seemed so far to make best, if imperfect, sense of our ethical questions generally? We must look to what we already know, or think we know, before we try to make sense of radically new ethical questions, because any radically new answers must build upon what is best about what we already know, or else patiently and persuasively show us why what we thought we most knew must be rejected.

However, we should have resisted the impulse, reasonable as it was, to attempt to develop environmental philosophy on any of the traditional theories or tempting amalgamations of them (something that Snyder himself, who is principally a poet, has not tried to do, even though what he *has* done presents us with a goldmine for reflection). We should have first faced up squarely to the fact that those theories themselves are one and all built primarily for human life and, thus, not according to any standard that we could assume was appropriate for evaluating the life of any other species that we know of.

What would it mean to resist the temptation to begin with one of the many theories that we are already familiar with, or some amalgamation of them? In philosophy nothing is new under the sun; we cannot expect to find an entirely new beginning, not echoing any fundamental insight of the past. And, anyway, if what we seek is an ethic, a moral philosophy, it must be in some sense already understood by everyone, however inspiring and enlightening it may seem to be when made explicit. Whatever is to guide and motivate us in our relationship to the rest of nature, it must be somehow familiar. Even so, we can look for something new relative to what seems to be contributing to our stuck-ness or scattered-ness of opinion. If possible, we should investigate first principles, doing our best to get as fresh a start as we can and at the deepest level possible.

Until recently, I do not think we (in particular, Western analytic philosophers) have been in a position to begin like this on the ethical questions raised by environmentalism or the conservation of nature. Certainly, there have not been any candidates for compelling first principles that might enjoy wide acceptance, principles that could unite and advance the diversity of opinion, and application of principles from traditional sources, Kantian, utilitarian, etc. But there is recent work on the logical form of the concept of life that, as we hope to show, derives the standard and grounds for any future environmentalism or ethics of the conservation of nature (if such is possible), at least insofar as that concerns the conservation of living things, in all their great variety of forms.

San Diego, CA Daniel Guevara

References

1. Cafaro, P. 2001. The naturalist's virtues. *Philosophy in the Contemporary World* 8(2):85–99.
2. Cafaro, P. 2001. Thoreau, Leopold, and Carson: Toward an environmental virtue ethics. *Environmental Ethics* 22: 3–17.
3. Hursthouse, R. 2007. Environmental virtue ethics. In *Environmental Ethics*, ed. R. L. Walker and P. J. Ivanhoe, 155–172. Oxford University Press.

4. Leopold, A. 1949. *A Sand County Almanac and Sketches Here and There.* New York: Oxford University Press.
5. Sandler, R. and Cafaro, P. (eds.) 2005. *Environmental Virtue Ethics.* Lanham, MD: Rowman and Littlefield Publishers.
6. Sandler, R.L. 2013. *Environmental Virtue Ethics.* https://onlinelibrary.wiley.com/doi/10.1002/9781444367072.wbiee090. Accessed March 11, 2023.
7. Taylor, P. 1986. *Respect for Nature: A Theory of Environmental Ethics.* Princeton University Press.
8. Payne, R. and McVay, S. 1971. Songs of Humpback Whales. *Science* 173(3997): 585–597.
9. IUCN Red List. 2022. https://www.iucnredlist.org/statistics. Accessed March 11, 2023.
10. Lorenz, K. 1974. *Civilized Man's Eight Deadly Sins.* New York: Harcourt Brace Jovanovich, Inc.
11. Wilson, E. 1984. *Biophilia.* Harvard University Press.
12. Sandler, R. 2012. *The Ethics of Species: An Introduction.* Cambridge: Cambridge University Press.
13. Safina, C. 2015. *Beyond Words: What Animals Think and Feel.* Henry Holt & Company Inc.
14. Sala, E. 2020. *The Nature of Nature: Why We Need the Wild.* Washington, DC: National Geographic.
15. Schaller, G.B. 1993. *The Last Pand*a. Chicago: The University of Chicago Press.
16. Attenborough, D. 2020. *A Life on Our Planet: My Witness Statement and a Vision for the Future*. Ebury Press.
17. Monbiot, G. 2017. *Out of the Wreckage. A New Politics for an Age of Crisis.* Verso Books.
18. Ward, D., Melbourne-Thomas, J., Pecl, G.T., Evans, K., Green, M., McCormack, P.C., Novaglio, C., Trebilco, R., Bax, N., Brasier, M.J. and Cavan, E.L. 2022. Safeguarding marine life: Conservation of biodiversity and ecosystems. *Reviews in Fish Biology and Fisheries* 32(1): 65–100.
19. Campagna, C., Guevara, D. and Le Boeuf, B. 2017. Sustainable development as *deus ex machina*. Biological Conservation 209: 54–61.
20. Snyder, G. 1980. *The Real Work: Interviews and Talks, 1964-79.* New York: New Directions.

Acknowledgments

We have been developing the ideas in this book for many years and are grateful for the support and critical engagement we have received during that time from Burney Le Boeuf, Jorge Hankamer, Giuseppe Notarbartolo di Sciara, Kyle Robertson, Julie Tannenbaum, and Bernd Würsig. Over the years, we also co-taught graduate seminars in the Department of Philosophy, and the Department of Ecology and Evolutionary Biology, at the University of California, Santa Cruz. There we had the opportunity to enrich our understanding, in discussion with the students, of many of the concepts in this work. To them our thanks. We also wish to express our gratitude to Paul Koch (Dean of Physical and Biology Sciences (2011–2022), and Distinguished Professor of Earth and Planetary Sciences, UC Santa Cruz), Peter Raimondi (Professor and (formerly) Chair, Ecology and Evolutionary Biology Department), and also to the late Tyler Stovall (Dean of Humanities (2014–2020) and Professor of History, UC Santa Cruz), for generous financial support and encouragement. We regret that Professor Stovall did not live to see this work.

We were also fortunate to be a branch of the Center for Public Philosophy at UC Santa Cruz (Professor Jonathan Ellis, Director), from its inception in 2014 until 2020. Our thanks to the Center for helping us establish ourselves at a crucial time, with a public platform. Thanks also to support from the Humanities Institute (formerly, the Institute for Humanities Research) at UC Santa Cruz (Professor Nathaniel Deutsch, Director, and Irene Polic, Managing Director), in particular for sponsoring and hosting three very successful interdisciplinary colloquia on our work. To our distinguished panelists in those colloquia—Professors Karen Barad, Paul Koch, Ronnie Lipschutz, Eric Porter, Daniel Press, and Beth Shapiro—our deepest appreciation.

The Wildlife Conservation Society generously supported Claudio Campagna's residence in Santa Cruz for one quarter per year, between 2009 and 2019. We thank John Robinson (Executive Vice President for Conservation and Science at WCS) and Guillermo Harris (Director of WCS Argentina Program) for their support. We are particularly grateful to Vienna Eleuteri (Eulabor Institute) and to Adam and Jessica Sweidan (Synchronicity Earth) for their support of the Montefortino workshop (Italy, 2017), where our ideas were so constructively tested and refined, and to Giuseppe Notarbartolo, Simon Stuart, and Peter Harris, who were crucial to the planning and execution of this very fruitful event.

We would like to thank our artists, for their beautiful and thought-provoking representations of main themes in our work: Eugenia Zavattieri, for the cover image, and Andrew Guevara, for the illustrations in the book itself. Many thanks to our patient and generous editors at Springer, Bibhuti Sharma and Éva Lőrinczi. And our deepest gratitude to Philip Cafaro, Cora Diamond, Carl Safina, Enric Sala, and Beth Shapiro for reading and commenting on the penultimate draft of the book, and especially for Cafaro's extensive comments on it. Finally, we thank our families, especially our wives—Adriana Guevara and Victoria Zavattieri—for their love and support, through the long and improbable adventure that resulted in this book.

Introductory Note

> Let the use of words teach you their meaning.
> L. Wittgenstein [1: §303]

> When I think in words, I don't have 'meanings' in my mind in addition to the verbal expressions; rather, language itself is the vehicle of thought.
> L. Wittgenstein [2: §329]

> Now what do the words of this language signify?—How is what they signify supposed to come out other than in the kind of use they have?
> L. Wittgenstein [2: §10]

The various practices of conservation differ in the language they use to represent the world they wish to conserve and this affects what they judge to be good or bad, acceptable or unacceptable, and, accordingly, judgments about what we should try to conserve and why. The language affects the very nature and significance of the practice. Our essay analyzes the language that conservationists use to represent and evaluate nature. Our analysis is especially concerned with how their language contributes to the crisis of species extinction, caused by routine and avoidable human choices and actions. We aim to demonstrate how that crisis, and any solution to it, has its origins in the language used to represent and evaluate what we wish to conserve.

Our analysis is inspired by what the philosopher Ludwig Wittgenstein calls "a grammatical investigation," which he explains this way:

> Our inquiry is therefore a grammatical one. And this inquiry sheds light on our problem by clearing misunderstandings away. Misunderstandings concerning the use of words, brought about, among other things, by certain analogies between the forms of expression in different regions of our language [2: §90].

"Grammar," in Wittgenstein's sense, is not principally about the rules we learn in grammar school for the inflection of verbs, or the like, e.g., the rule for whether a verb in English takes "ed" or not in the past tense. The science of modern linguistics came after Wittgenstein's day, but its formal investigations in semantics and pragmatics are closer to what he intends by his deeper sense of "grammar" (see [3] and Chap. 10), exemplified by the logic that guides our language and thought. The crisis of species extinction begins with difficulties in the language used to represent and understand life. These are difficulties that can only be appreciated and resolved with attention to this deeper sense of "grammar." Throughout our book, we develop "the language of the forms of life," which best illustrates and confirms how a grammatical investigation is crucial to all conservation practice.

References

1. Wittgenstein, L. 2009. *Philosophy of Psychology - A Fragment.* (trans: Anscombe, G.E.M., Hacker, P.M.S. and Schulte, J. Revised fourth edition by Hacker P.M.S. and Schulte J.). West Sussex: Wiley-Blackwell.
2. Wittgenstein, L. 2009. *Philosophical Investigations.* (trans: Anscombe, G.E.M., Hacker, P.M.S. and Schulte, J. Revised fourth edition by Hacker P.M.S. and Schulte J.). West Sussex: Wiley-Blackwell.
3. Carlson, G.N. and Pelletier, F.J. 1985. *The Generics Book.* Chicago: University of Chicago Press.

Contents

About the Authors

Claudio Campagna has an M.D. from the University of Buenos Aires, Argentina, and a Ph.D. in Biology from the University of California, Santa Cruz. He started his career as a scientist working on the behavioral ecology of marine mammals. But the core of his career was spent as a conservation biologist with the Wildlife Conservation Society. He has published widely in the fields of animal behavior and conservation, including *Ethology and Behavioral Ecology of the Otariids and the Odobenid* (co-edited with Robert Harcourt; Springer, 2021). In the field of conservation and language, he has published *Bailando en Tierra de Nadie: Hacia un Nuevo Discurso del Ambientalismo* (Editorial del Nuevo Extremo, 2013).

Daniel Guevara is Professor Emeritus at the University of California, Santa Cruz. He did graduate work in philosophy at Princeton University and the University of California, Los Angeles, taking his Ph.D. from the latter. While at UC Santa Cruz he taught widely in ethics and the history of philosophy, including environmental ethics, moral psychology, Kant, and Wittgenstein. He has published widely in these fields as well, including *Kant's Theory of Moral Motivation* (Westview Press, 2000) and *Wittgenstein and the Philosophy of Mind* (Oxford University Press, 2012), co-edited with Jonathan Ellis.

1 Uncovering Grammars

I have always honoured those who defend grammar and logic and it is only realised fifty years later that they have averted great dangers.
M. Proust [1, §123]
If you want to act like a robot, how does your behavior deviate from our ordinary behavior? By the fact that our ordinary movements cannot even approximately be described by means of geometrical concepts.
L. Wittgenstein [2, §324]

At the Tokyo Olympics, in 2021, the synchronized swimming team from the USA performed a piece in which the athletes moved as though they were robots [3]. Their movements could have been described by concepts of geometry: angles and bisections, straight and parallel lines, compass turns. It is mildly paradoxical that the most expressive and graceful human movements we can imagine should lend themselves to representation in language inadequate for ordinary human behavior. The geometric terms which describe the robotic movements of the swimmers are not those which describe the flexibility, agility, and subtlety of the movements we ordinarily find in the human form of life. Attempts to describe ordinary human behavior in robotic terms would, on the contrary, tend to involve the attribution of grave defects, like those caused by neurological disease or injury. If the language of ordinary human behavior was all we could use, then the Olympic choreographers would have presented us with a real paradox: representing superb human movement as defective.

Robotic movements require effort for human beings; coordinating them in a group performance was, of course, part of the power of the choreographed piece. There was no serious risk of anyone mistaking it for defect; synchronized swimming can be presented mechanically, like perfectly synchronized gears. Any idea of badness or defect would derive from a misunderstanding of the performance.

C. Campagna, D. Guevara, *Speaking of Forms of Life*, Fascinating Life Sciences,
https://doi.org/10.1007/978-3-031-34534-0_1

A misunderstanding like this commonly occurs in the practice of conservation because the prevailing language used to address the threat of human-induced species extinctions is not suitable for the correct representation of the life forms being extinguished. As a consequence, what is wrong, what is lost, when we drive a species to extinction is misjudged. The resulting error is grave because it leads to interventions that may be irrelevant, superficial, weak, and inadequate to the task of addressing the full significance of the extinctions we provoke. A language whose grammar guides the expressions of geometry cannot represent, not even approximately, ordinary human movement. And, as we shall argue, a language whose grammar guides our representation of species in terms of how they are 'useful' to human beings, or in terms of their 'beauty,' or as populations to be 'sustained,' or in any of the terms that now dominate the language of conservation, cannot even approximate the harm done by the extinctions we unnecessarily cause.

If there were no known alternative to the language of geometry, or if we obstinately adhered to it in trying to describe our ordinary movements, our attempts would betray a mismatch that could be remedied only by finding a language whose logic made sense of concepts like elasticity and fluidity. There is a similar difficulty that occurs when, for example, the language of 'sustainable development' tries to represent what is lost when human beings unnecessarily drive species to extinction. But the difficulty is covered up by terms borrowed from another region of our language, the language of economics, while the ineptness of the grammar goes unnoticed, and so our inept practices proceed undisturbed.

Human-Caused Extinctions

It is the extinctions *we 'provoke'* that are of particular importance in this essay (Chap. 4, Box 4.1). The term differentiates a particular mode of extinction, distinct from the kind that occurs in nature without human participation. Natural catastrophes can cause extinctions. Such causes are essentially different from our over-exploitation and destruction of nature. The extinctions we provoke, unlike those that form part of the evolutionary history of a species, are a force of our will, induced by our voluntary participation. Practitioners of conservation understand this difference and yet often represent it with a language that homogenizes it, treating it as though it were one and the same thing—a prime example of a fateful flouting of grammar.

Only those extinctions caused by avoidable human behavior raise the issue of what can and ought to be corrected; there is nothing a catastrophe *ought to* have done. Then, the specific language used to describe the extinctions we provoke, and the threats we unnecessarily pose, must be distinguished from that which describes an extinction's natural role in evolution, independently of human choices. The grammar of the language of provoked extinctions should be, fundamentally, one that can represent what living things themselves are, independently of their utility or aesthetics, or any other criteria that may have nothing to do with how they flourish. The tacit constraints and implications of such a grammar would necessarily differ from that which is used to describe extinctions due to natural causes, much as the logic of geometrical descriptions of robotic movements must differ from that which underlies our ordinary descriptions of human bodily movements. Ignorance of this kind of difference in language, willful or otherwise, is at the root of many of the crises of conservation. It is especially detrimental to the conservation of species because it contributes to the most fundamental misunderstanding of all, namely, a failure to understand what natural life itself is.

"A Great Moral Wrong"

The extinction events which predate the human species are well described by the language of evolutionary biology, geology, and paleontology, among others. On the other hand, *provoked* extinctions must be represented by a different kind of language. It has been said, for example, that "a [provoked] species extinction is a great moral wrong" [4, 5]. A judgment like this exceeds what is possible with the grammar of evolution, or geology, etc., inasmuch as the latter involve no moral concepts. In contrast, it seems perfectly reasonable to raise moral issues about extinctions induced by over-exploitation or other instances of human-caused extinction. Not restraining our use of nature, while knowing the problematic consequences of unrestrained use, raises such issues and may lead to judgments which introduce the idea of a moral need to redress what is judged as wrong or bad, or to the judgment that there is something morally wrong with us when we do nothing to intervene in the face of our problematic behavior. A moral need, and especially a need to address evil (e.g., slavery), introduces the idea of something being unconditionally bad or wrong, and of an unconditional imperative to change it. Contrast this with etiquette or, even the law, where there is not necessarily any comparable, unconditional imperative implied: sometimes it is ethical to be impolite, or even to break the law (against slavery, for example). With morality, or moral evil at least, our evaluative judgments are often intended as unconditional.

Are provoked species extinctions a crime comparable to the evils of torture or slavery [5]? Is the language which condemns the killing that drove the passenger pigeon to extinction comparable to the condemnation of genocide? If so, we might ask why the crisis of human-induced extinction is so far from being resolved, and in fact only worsens. Our contention in this essay is that we must learn how to extend our moral grammar in this direction before we can answer such questions.

Pluralism and Commensurability

The catastrophic disappearance of the dinosaurs and the annihilation of the passenger pigeon differ in such essential ways that it is remarkable how they are often treated as though they fall under one and the same concept of extinction. But neither does it seem right to assume that we can evaluate the pigeon's provoked extinction with the moral language that represents evils between humans. Though some are willing to condemn it morally, conservationists still debate whether in fact moral condemnation can apply to provoked extinctions. And even when they agree that human-caused extinctions are bad, they often disagree on the standard used to judge what is good or bad. Those who judge an extinction to be morally bad or wrong do not necessarily judge according to the same standard as those who judge the badness in terms of the destruction of a 'natural resource' or the destruction of something whose beauty we enjoy. Considering the variety of standards, some conservationists embrace a pluralistic approach that tries to honor all the various points of view (Chap. 16). However, as we will see, this just magnifies the problem by dispersing it across a multitude of evaluative terms, assumed to be commensurable, when in fact the grammar that guides our use of the terms is following quite different standards: aesthetic, economic, etc., which cannot necessarily be weighed and balanced on the same scale of measure. To assume that the plurality of evaluations are commensurable is like assuming that all the language we might use to describe human movement, including the language of geometry, is interchangeable.

Therefore, our essay questions the assumption of commensurability, or homogeneity, in the language that conservationists commonly use to represent and evaluate provoked extinctions. Nor are we in accord with those who see advantages in the plurality of representations. The plurality of standards has divided conservationists into factions, often speaking past each other in a kind of conservationist Babel. A consequence of pluralism is that what is condemnable or permissible varies relative to each standard, even though the same objective facts are being evaluated. It seems obvious, however, that the species in danger of extinction ought itself to be the focus of the evaluation of its extinction as bad or wrong, without having to accommodate, on an equal footing, values that vary across other standards having nothing to do with what it is to be the relevant kind of life.

Echoing the passage from Wittgenstein, cited above (Introductory note), we can say that plurality in the language used for evaluating extinctions leads to misunderstandings prompted by, among other things, certain similarities between the forms of expression in various "regions of our language." Similarities between descriptions of provoked extinctions and those involving no human participation can lead to certain misunderstandings. For example, in light of the fact that sooner or later all species go extinct, the particular harm caused by our voluntary choices might seem not to matter. This shows that we need to represent provoked extinctions with a language that makes our threat to life itself the focus of evaluation and implicates human will in what is judged bad about the loss of a life form.

Confronted by the various standards of evaluation, the various currents in conservation fail to converge upon the fundamental reason that might be used to ground what is bad or wrong in our causing an extinction: namely, the unnecessary and permanent loss of a life form and all the value of its flourishing. Instead, what is 'bad' might variously be taken to lie in the fact that extinction of a 'top' predator affects the trophic structure, or that, because the species has 'disappeared,' we can no longer experience its beauty, or buy and sell it, or consume it.

'Administered Extinctions'

The lack of a proper language for understanding life forms, and for agreement on a unifying standard for evaluating provoked extinctions, results in judging extinctions as inevitable or even as good. It may be 'good' ('desirable,' 'acceptable') that 'harmful' or 'dangerous' species be extinguished, as was once likely said of the passenger pigeon, extinguished more than a century ago. But the most fearful consequence of a plurality of standards is that it opens the door to certain strategies of extinction, as suggested by this:

> No one cares for the prospect of consigning threatened species to oblivion. But insofar as choices are already being made, unwittingly, they should be made with selective discretion that takes into account the impact of the extinction of a species upon the biosphere or on the integrity of a given ecosystem [6, Ch. 6, §67].

The passage is taken from the Brundtland Report, the document which installed the concept of sustainable development and that has served as the foundational document for world-class conservation. The report is proposing that provoked extinctions be managed, by accepting some over others, for a supposed overall good balance of a plurality of values. The absurdities of language are plainly in view here: consider how "the choices are already being taken, involuntarily" represents the annihilation of species as unintentional, as though all extinctions were inevitable natural phenomena, occurring whether we like it or not. The administrative strategy appears to encourage some kind of rational selection based on instrumental values, as is even more clearly stated in another passage from the same document:

> The genetic landscape is constantly changing through evolutionary processes, and there is more variability than can be expected to be protected by explicit government programmes. So in terms of genetic conservation, governments must be selective, and ask which gene reservoirs most merit a public involvement in protective measures [6, Ch. 6, §10].

Both passages advise the world's governments to use discretion in managing the variety and variability of life, as though compromise were inescapable and not caused by our voluntary choices. The language sets up a pragmatic framework for treating living things as instruments to a bureaucratic end. The framework values the threatened forms of life only as they affect the overall balance of costs and benefits while ignoring the role of avoidable human actions. Which species is saved, and which sacrificed, is justified by what the species 'deserves,' as measured by human interests.

Any act of conservation intended to rectify the wrongs that flow from human-induced threats of extinction is influenced by the way it represents and evaluates life and living things. If the representation is disconnected from what is necessary for life and a creature's flourishing, then it cannot address the wrong of extinguishing the forms of life, much as the language of geometry cannot describe or correct ordinary human movement. Our essay maintains that the failure to properly represent life is a cause of the crisis of conservation and contributes to its deepening. The language of governments does not provide incentives to stop provoked extinctions, but settles for mitigating them at best, and even that goal is presented as possibly unfeasible. The passage just quoted from the Brundtland Report [6, Ch. 6, §10] starts with this:

> Some genetic variability inevitably will be lost, but all species should be safeguarded to the extent that it is technically, economically, and politically feasible.

There are then preconditions, instrumental requirements, to be satisfied before provoked extinctions are stopped. This conclusion, we will argue, is unjustifiable, if our primary concern is the life forms themselves and the creatures that bear them. We propose to describe, analyze, and develop a language whose grammar departs dramatically from the Brundtland Report, and justifies interventions to stop provoked extinctions independently of technical, economic, or political compromises. This is the language whose grammar makes possible the representation of life in the first place, and which serves, at the same time, as the standard for evaluating what is wrong when we drive species to extinction. And it is the language that serves as the grounds for motivating whatever interventions we might implement to protect them.

Language as Intervention

The language we propose and develop relates human choices to what living things need in order to satisfy the necessities of their life forms. In order to show how this language can sustain conservation practice, we need recourse to ideas that originate in the philosophy of language, the theory of value, and natural history.

When Wittgenstein asks "What is your aim in philosophy?" he responds with "To show the fly the way out of the fly-bottle" [7, §309]. Obviously, the fly is not a form of life that can receive instructions for how to escape, but the human is. We wind up trapped in the bottle through the misuse of language, and, particularly, through our evaluations of whether an extinction we provoke is wrong or bad. Attention to language can also show us the way of escape.

The common practices of conservation, as we hope to make clear throughout this book, largely ignore language as part of the problem of, or the solution to, the crisis of conservation. They tend to limit language to its role as a tool for communication, for example, instead of recognizing it, as Wittgenstein does, as the vehicle of thought itself [7, §329] and, thus, of our apprehension of reality. Questions about the choice of language are commonly about rhetoric or vocabulary; it is often assumed that a simple change of words is all that is needed to clarify misunderstanding. The rules behind the construction of an expression, its 'architecture,' in the sense of the grammatical structure that guides the assembly of its words into a meaningful expression, are rarely if ever considered as key to an intervention for redressing threats to species. Instead, the typical hope is that the crisis of provoked extinctions will be remedied by science, or a legal victory, or administrative decisions. The question of whether human actions are completely intolerable and unjustifiable when they unnecessarily interfere with what the other forms of life need to develop and flourish is not among the fundamental considerations guiding conservation; it never has been. At best, certain conservationists naively take it for granted that it is an important consideration, with no concern for verifying this explicitly. This way of proceeding impedes attention to the grammar behind the way conservation describes the state of the world and frames all its resulting evaluative judgments.

In our ordinary use of language—in representing and evaluating ordinary human movement, for example—we do not need to explicitly attend to or even identify the logic behind our language—no more than we have to attend to the physics of motion to walk. Ordinarily, the fly flies freely, not trapped in the bottle. In important contrast, the practice of conservation *does* require such attention to language, in order to see why so little progress is made and why we keep running into limits that we cannot see until we hit them. It requires careful attention to what Wittgenstein calls "depth grammar," or, usually, just "grammar." [1] These expressions concern precisely the rules or logic that makes meaningful language use possible. What is important for practice is the architecture, the form and structure of this grammar.

We emphasize: This is not 'grammar' in the relatively superficial sense of how to conjugate a verb—e.g., "conserve"—in various languages, but rather in the sense of how the term is used to install a particular concept, including its implications for our values, the ends that guide our practice. The iconic term, 'sustainable development,' has installed the instrumental value of the rest of nature as a primary standard for judging the interventions thought necessary to conserve what is good and to correct what is wrong in the world. By an artifice of the grammar of this language, nature's instrumental value, its value as a service to us, turns out to be a primary reason for practicing nature protection, at least at the governmental level, and a guide for its interventions, all without clear and stable connection to what a living thing needs to flourish or to the very thing that the lack of 'sustainability' supposedly threatens. Occasionally, one finds alternative talk of the 'intrinsic value' of life. But the term does not clarify what it is to be a living thing, much less what it needs to flourish (Chap. 13).

To summarize: whoever practices conservation for aesthetic reasons, for example, practices a conservation distinct from one which is based on the value of species relative to the services they offer. The language of each facilitates, impedes, induces, evades, and obligates the practice. The expressions which emerge from each current in conservation follow the shape of the logical grammar that makes them possible, like a river that follows the mold of the riverbed. With talk of lines and angles we

[1] He says: "In the use of words, one might distinguish 'surface grammar' from 'depth grammar.' What immediately impresses itself upon us about the use of a word is the way it is used in the sentence structure, the part of its use—one might say—that can be taken in by the ear.—And now compare the depth grammar, say of the verb "to mean", with what its surface grammar would lead us to presume. No wonder one finds it difficult to know one's way about." [7, §664]

cannot comprehend, or even register, the subtleties of fluid movement. And whoever understands movement only in terms of a fluid dance will have difficulties imagining a choreography that suggests the behavior of a robot. A practice which expects to identify and correct the wrongs of provoked extinctions must examine the language it uses to declare what it understands by "wrong" or "bad," with a view to uncovering the grammar that guides it.

Conclusion

Wittgenstein speaks of a "grammatical investigation" as one that can remove misunderstandings. The chapters that follow will clarify what this means, by clarifying what a grammatical investigation is, in the context of the conservation of living things. Its importance lies in an investigation of the use of words, for their definition or meaning is found there. In the effort to clarify the many misunderstandings that surround our talk of provoked extinctions, we will uncover and examine the unique grammar that is essential to the language we use to represent life, and the standard implicit in that language, for evaluating whether a living thing is doing well or badly. We call the practice which emerges "the conservation of the forms of life." This practice is an intervention which begins fundamentally with attention to the logic of our language, in order to eliminate the many misunderstandings that arise from the many misrepresentations of life—including those found in biology—and that prevent us from comprehending what is lost when we unnecessarily drive a species to extinction.

We draw on the work of philosophers who have shown what the underlying grammar is of the proper representation of life, the one without which it would be impossible for us to so much as identify living things in the first place and to distinguish them from the non-living. And our ultimate purpose is to show that this representation also contains the primary standard for evaluating as good or bad, right or wrong, anything we do that affects the other life forms.

Our critique of any conservation based on other ways of representing life, or its provoked extinction, is that all resulting interventions and remedies have lost *logical* contact with the proper understanding of life. We will see that plenty of conservation actions seem to be driven by the purpose of the flourishing of the focal, living individuals, and yet, under careful scrutiny, it is not obvious that the standard of flourishing is that of satisfying the needs of the form of life. If interventions on behalf of the other life forms—and attempts to conserve what we judge to be good or to remedy what we judge bad—lose the connection to the logic of the language that represents life, then they will also lose any connection to the flourishing of those life forms. They might as well be interventions based on a language that describes robots.

Box 1.1: Language

> Language is a labyrinth of paths. You approach from *one* side and know your way about; you approach the same place from another side and no longer know your way about.
>
> L. Wittgenstein [7, §203].

We use language to ask for help, give orders, point to objects, express desires or preoccupations, guide a process, greet someone, congratulate or criticize them, among countless other things. Wittgenstein analogizes the use of language to the playing of a game (Box 1.2). The use of language involves the capacity to put together intelligible expressions that accomplish all we do with language, in ways comparable to the capacity for learning how to play all the various kinds of games there are. Asking for help, giving orders, making promises, etc., are moves in the various 'games' of language that we 'play.'

Of utmost importance to us in this essay is the language we use to talk about life, especially the use we make of it when we describe living things and evaluate them based on those descriptions. We maintain that mitigating the threat of human-induced extinction, affecting so many life forms, depends on this language.

This essay proposes, develops, and defends a particular language-game as crucial to the understanding of life and its conservation. It is a language familiar to the practitioners of conservation, namely, the language of natural history. Certain naturalistic expressions that we later call "the language of the forms of life (or life forms)" describe living things according to a specific and unique grammar. We maintain that there is a standard implicit in this grammar that grounds the primary reasons for redressing the threat of human-induced species extinction. Indeed, we argue that it grounds the primary reasons for the practice of conservation generally if what we primarily care about are the living things themselves. Our contention is that the grammar implicitly expresses the necessities of the forms of life themselves and that it is only these necessities that could justify any reasons we might have for intervening against, and condemning, the threat of human-induced extinctions as categorically unacceptable. As we hope to show, it is a sad irony that the language-games that dominate, and have dominated conservation, have instead led down the labyrinth of many other paths and lost the crucial connection to the necessities of the life forms.

Box 1.2: Language-Games and Grammar

Although they are key terms for him, Wittgenstein never defines "language-game" or "grammar." He expects us to discern his meaning from his use. Our discussion is inspired mainly by his usage in the *Philosophical Investigations* [7]. We do not pretend to offer a scholarly or systematic treatment, but hope that our own meaning is discernible from our use.

Language-games

A main purpose of the analogy between language and games in the *Investigations* is to correct for certain common misconceptions about language: e.g., that it is mainly for communication or for representing what is true of false. It has those functions, of course, but when we focus on them, to the exclusion of others, we misunderstand or miss altogether the great variety of things we do with language, as many of Wittgenstein's examples of language-games show. For example, he lists, among other things [7, §23]:

- Giving orders and acting on them
- Making up a story; and reading one
- Acting in a play
- Cracking a joke; telling one
- Forming and testing a hypothesis
- Requesting, thanking, cursing, greeting, praying

Wittgenstein also says there that he uses "language-game" to emphasize the fact that "the *speaking* of language is part of an activity, or of a form of life," [7, § 23] and that,

> We don't notice the enormous variety of all the everyday language-games, because the clothing of our language makes them all alike. What is new (spontaneous, 'specific') is always a language-game [8, §335].

The word "game" itself is a piece of 'clothing' that might cover up the diverse variety of activities it denotes. It might, for example, suggest mere diversion or entertainment, but Wittgenstein's examples belie this suggestion. Like the playing of games, the use of language is sometimes highly rule-bound and goal-directed (like chess or tennis); at other times, it is marked by

(continued)

Box 1.2 (continued)
playfulness and capriciousness. A language-game is sustained by the words and the actions with which it is interwoven, including the gestures, the facial expressions, the tone with which the words are accompanied, and more generally, our interests and sensibilities, what strikes us as important, funny, similar, etc.

We will see that there are many language-games in conservation—collecting data, forming of hypotheses, declaring threats to species, finding reasons to intervene on their behalf, with all consequences that may follow. These are often covered up by words or phrases (e.g., "sustainable development") like clothes that make them appear tidy, well-meaning, and effective, when they are not. Our use of language—its rules and its vicissitudes—like the games we play only makes sense in the context of human life. And the language of the conservation of species only makes sense in relation to what it is to be another form of life.

Grammar

The grammar we are taught in school, or when learning a foreign language, is comparable to a game where there are explicitly stated rules, like those about how to move certain chess pieces or how to set them up on the board. These are the rules for learning how to form past or present tense for example, or how to start or end a sentence, etc. Wittgenstein is concerned with a deeper sense of grammar, one that is not arbitrary or mere convention, but can reveal the logic of one's language. The best example of grammar in this sense is the one central to this essay, explained especially in Chaps. 10 and 11, and Boxes 10.1 and 12.1, where we lay out the logic of the language-game that represents life. For all the differences between the grammar we learn in grammar school and the grammar that interests Wittgenstein—and that informs the central themes of the essay before you—we, ordinary speakers, are the masters and authoritative source of both. We understand, for example, the logical grammar of "life," "living thing," "life form," by looking to our use, as competent speakers of the language. In this way, "[e]ssence is expressed in grammar," as Wittgenstein says [7: §371]: i.e., we know what life and life forms are because we know how to describe them according to the grammar that guides our use.

©Andrew Guevara

Box 1.3 The Language of Representation and the Language of Evaluation

Our book depends crucially on the proper representation of life by means of language. Following philosopher Michael Thompson [9, 10], we maintain that this is found in the descriptions typical of the natural history of a life form. In a textbook on bees, "the bee is an insect," "the bee has wings" and "the bee

(continued)

Box 1.3 (continued)

brings nectar to the hive," all provide knowledge of the natural history of the bee. They all express what the bee, as a life form, *is, has, or does.* They describe necessary truths about the kind of life form it is.

In "the bee is useful because it pollinates crops," knowledge is conveyed as well, but not a necessary truth in the natural history of the form of life. Rather, the proposition evaluates bees by the service they provide to others—humans, presumably. This is also true of "the bee is dangerous," or "the bee has pretty colors." To learn that the bee is dangerous is to learn something of instrumental value, that the bee is pretty communicates information about its aesthetic value, based on taste. It is important for conservation practice to keep track of when it introduces some evaluative component like this into the representation of a life form, and not to assume that it says anything necessarily about the form, as opposed to what it says about us, our interests or attitudes.

The rationale for a conservation practice is shown in the language it uses to represent and evaluate living things. This can be seen by breaking practices down into essential stages: representation or description of living things, some evaluation of this, and intervention based on these two previous stages. The language of natural history describes life forms by describing the necessities of the forms, and those necessities also implicitly involve a standard we can use for evaluating individual creatures as doing well or badly, insofar as they satisfy, or not, the necessities of their respective forms; representation and evaluation are therefore intimately and inseparably linked here (see Chaps. 8 and 9). In contrast, judgments about the usefulness to us of a life form, or about its aesthetic qualities, attach an evaluation based on a standard not necessarily related to the form. The language of these other judgements is often intermingled with natural historical language, particularly aesthetic language. These other judgements are important in the practice of conservation, but for reasons unrelated to the natural historical necessities dictated by the form.

In so far as conservation interventions depend on such representations and evaluations, conservation practice is utterly dependent upon the language it uses to represent and evaluate the objects of its concern. *The logic of the language of representation and evaluation is critical to the practice. Leaps in logic generate leaps in value systems, and, therefore, in the reasons for practicing conservation.* Conservation practices are not usually preoccupied with such matters and, as a result, are often unaware of how far removed their reason for being has strayed from the primary concern for life itself.

References

1. Proust, M. 1927. *Time Regained* (*In Search of Lost Time*). Vol. 7 (trans: Hudson, S.). https://gutenberg.net.au/ebooks03/0300691h.html. Accessed 11 Mar 2023.
2. Wittgenstein, L. 1980. *Remarks on the Philosophy of Psychology*. Vol. I (trans: Anscombe, G.E.M.), ed. G.E.M. Anscombe and G.H. von Wright. Chicago: The University of Chicago Press; Oxford: Basil Blackwell.
3. https://www.facebook.com/watch/?v=492168118760581. Accessed 11 Mar 2023.
4. Cafaro, P., and R. Primack. 2014. Species extinction is a great moral wrong. *Biological Conservation* 170: 1–2.
5. Cafaro, P. 2015. Three ways to think about the sixth mass extinction. *Biological Conservation* 192: 387–393.
6. Report of the World Commission on Environment and Development: *Our Common Future* (Brundtland Report). 1967. Transmitted to the General Assembly as an Annex to document A/42/427—Development and International Co-operation: Environment. http://www.ask-force.org/web/Sustainability/Brundtland-Our-Common-Future-1987-2008.pdf. Accessed 11 Mar 2023.
7. Wittgenstein, L. 2009. *Philosophical Investigations* (trans: Anscombe, G.E.M., Hacker, P.M.S., and Schulte, J. Revised fourth edition by Hacker P.M.S., and Schulte J.). West Sussex: Wiley-Blackwell.
8. ———. 2009. *Philosophy of Psychology—A Fragment* (trans: Anscombe, G.E.M., Hacker, P.M.S., and Schulte, J. Revised fourth edition by Hacker P.M.S., and Schulte J.). West Sussex: Wiley-Blackwell.
9. Thompson, M. 2008. Life and Action: Elementary Structures of Practice and Practical Thought. Cambridge: Harvard University Press. See also: Thompson, M. 1998. The Representation of Life. In Reasons and Virtues: Philippa Foot and Moral Theory. ed. Hursthouse, R., Lawrence, G. and Quinn, W. Oxford: Clarendon Press.
10. Thompson, M. 2004. Apprehending Human Form. Royal Institute of Philosophy Supplement 54: 47-74.

2 About the Authors

The Intuitive Stage

Around 2008, we were brought together by discontent. We had both lost confidence in 'environmentalism' as a solution to the threat of provoked extinction, affecting so many species. The threat was on the rise, over-exploitation and destruction of habitats overwhelmed natural systems, human population growth had not ceased, consumption even less so. Obviously, something was wrong and it was seen by others (e.g., [1–3]); something was lacking: too much focus on our own species, too little on others, too much 'cost-benefit analysis,' too much ungrounded optimism, too little uncompromising opposition to the unacceptable, too much giving in, too conciliatory an attitude, too little self-criticism, too little institutional criticism. Interventions to prevent extinctions focused on the passing of laws, gathering and presenting scientific data, and reading the political climate. Laws were evaded; data did not change prevailing choices and attitudes. Our best efforts failed us. Conservation had lost touch with categorical values: indisputable, non-negotiable values. Its efforts to solve the crisis of provoked extinctions only deepened it, and in the process, principles were compromised and dysfunctional models continued to entrench themselves.

* * *

"Café Müller" is a dance piece of the choreographer Pina Bausch [4]. In a room packed with chairs scattered about, a woman walks around with her eyes closed. She never trips over the chairs because someone else in the scene moves them out of her way. Fragile and vulnerable, she depends on the clearing of her way, possibly never knowing it or noticing it. This piece suggests an ironic analogy with the practice of conservation: conservationists deploy their interventions with confidence in the objectivity of the natural sciences and with the pragmatism of economists to guide them, but, even with these 'eyes,' they have not avoided colliding into many obstacles.

C. Campagna, D. Guevara, *Speaking of Forms of Life*, Fascinating Life Sciences,
https://doi.org/10.1007/978-3-031-34534-0_2

It was in this context of hidden dangers that we found, in the *Philosophical Investigations* of Ludwig Wittgenstein, and other works of his, ideas foundational for a study dedicated to the language of conservation. The discourse common to conservation, and especially that meant to address the crisis of species extinction, appeared to us to produce the sort of spell on the understanding that Wittgenstein speaks of when he says, "Philosophy is a struggle against the bewitchment of our understanding by means of our language" [5, §109, translation amended]. The objective language of science evokes a kind of enchanting confidence, but it is not primarily concerned with expressing the essential and grave significance of our provoking an extinction through our routine and avoidable behavior. For that we need a different language, one prepared to say what is good or bad. For that, conservation has turned, for example, to the language of economics, for which extinction may be a loss of an opportunity for 'development,' a loss of 'natural capital' or 'natural services.' Or aesthetic language has been used to express the value of meeting the human need for experiencing the beauty of nature. Both types of value are based on standards that are not essentially concerned with necessities of the life threatened by extinction; they have led us to treat the value of living things as lying in the satisfaction of interests or needs other than those of what a creature needs to live and flourish. Finally, environmentalist have introduced talk of 'intrinsic value' as a counter to these tendencies, but in practice this talk has been comparatively unpersuasive and not resulted in lasting change[1] (see Chaps. 13 and 25). The various perspectives most common in conservation today—with their emphasis on use value, aesthetic value, and intrinsic value, among others—differ in their language of representation and evaluation, but none of their interventions have corrected the course of annihilation. They may have deepened it.

The book *Bailando en Tierra de Nadie (Dancing in No Man's Land: Towards a New Environmental Discourse* [6]) pertains to around the time the authors met. One finds there: "When language speaks, there is no refuge in confusion because it functions like pointing with one's finger." The expression "language speaks" refers to when language is used as a self-evidently authoritative representation, from which we can derive values that, for example, ground the incontrovertible need to confront and eliminate a threat. *Bailando* proposes that conservation establish itself on a language that is like the pointing of a finger indicating what is obviously there. At the time we did not have the philosophical foundations to support the proposal. The book before you aims to resolve what was left pending in *Bailando*.

[1]The Endangered Species Act of the USA, insofar as it was established out of concern for the value of species for their own sake, was a significant victory of proponents of the intrinsic value of nature, even if in fact that was not the only value the Act addressed. But it is also a good example of how our best efforts in conservation have eventually failed us, the law having been avoided and eroded relentlessly over the decades, in part, we would suggest, because of vagueness around the concept of intrinsic value. Further in the book, we will suggest that the practice of life form conservation need not oppose what may intuitively be understood by 'intrinsic value'; indeed supporters and users of the concept will find in the language of natural history the grammar that is required to articulate it in rigorous detail.

Stage of Discovery

Wittgenstein put language under a magnifying glass. Perhaps, his approach would have practical application in unexpected places, like the conservation of species. We set out to explore the possibility that, in its efforts to address the havoc threatened by provoked extinctions, conservation was thwarted by the very language it used to represent the threat because the language was incapable of expressing the values necessary to comprehend it. As with other 'environmentalists,' experience had accustomed us to the various uses of environmentalist language. We had become accustomed to advancing arguments with the language of scientific data and that of the legal system, for example; we had pursued the ideas of 'sustainable development' and 'ecological economics.' We had learned from the twentieth-century pioneers of environmentalism and conservation—John Muir, Aldo Leopold, Rachel Carson, David Brower—and Henry David Thoreau, from the century before them. We identified with their aesthetic sense and their ideal of 'pristine' nature. Looking back now, we see that Leopold had glimpsed the heart of the matter, when, in discussing the other life forms, he speaks of "values as yet uncaptured by language" [7, p. 96]. But it was Wittgenstein who provided real understanding of how the problems of conservation were due to the inadequacy of our language to express those values—particularly, the loss of extinction.

Wittgenstein directed our attention to the architecture of language, to the logical structure that makes possible a meaningful assembly of words and discourages, or deactivates, other constructions as 'ungrammatical' in his sense, and thus meaningless. Language can lead the understanding through various ways of valuing things, including those where the primary value is to be found in what a living thing is, has, or does. But, as we will argue, it is only in this primary value that the fundamental reasons for practicing conservation lie in the first place, uncompromised by the interference of other language-games. The practice of conservation, as we came to understand, must pay attention to its use of its expressions of value, the 'moves' in its evaluative 'language-games.' The strengths or weaknesses in its representations of provoked extinctions have roots in those expressions.

Another choreographed piece of Pina Bausch, "*Tied Down*" [8], illustrates metaphorically these discoveries. In a minimalist scene, an empty, ample room opens out to contiguous spaces that one can only glimpse. In that room, a woman finds her movements limited by a rope attached to the opposing wall and holding her from the waste. The length of the rope determines the limits of her movements around the room. They are impassable limits; all effort to overcome them is in vain. And although her hands are free, she does not untie herself; it is as though she is not aware of what holds her. Language functions like that rope. It has constraints that limit the 'moves' we make in the language. The 'space' for moves permitted in the game depends on its 'rope,' its 'ties,' its 'grammar.'

These analogies help us understand how language impedes, channels, re-routes, clarifies, and (as Wittgenstein says) 'bewitches.' The language of sustainable development, for example, expresses value in economic terms. The underlying rules of this language-game, its rope, determine the possibilities of meaningful discursive movement. The game advances only as its 'rope' permits, firmly held by the ties that bind it. Sustainable development permits the representation of living things as useful resources and the derivation of evaluative judgments about them on this basis. Talk of 'appropriating' them is, therefore, an allowed move. Similar things can be said of the language used in the various other ways of practicing conservation: the language of species as 'populations,' or of their aesthetic or spiritual qualities, or of their 'welfare' or 'intrinsic value.' The corresponding practices all have their evaluative judgments derived from the terms in which living things are represented. Terms that are interwoven, according to a logic that determines what is a permitted move in the game. Each practice advances only as its rope allows; each conservationist practices her particular dance.

The Foundational Stage

By 2011, we had formed a few useful hypotheses and metaphors. We understood that the prevailing discourses had ropes that were far too short, or alternatively, ropes that allowed access to discursive spaces that were useless for eliminating the causes of the crisis of provoked extinctions. Untying ropes meant questioning the prevailing language of representation and evaluation. This hardly looked important to anyone practicing conservation at the time, nor has this changed! Each practice has perfected its dance and there seems to be no need to explore new spaces of action. If there is an underlying problem of the philosophical kind, it seems to have no effect on pragmatic concerns. '*Doing*' conservation is more important than proper grounding of the movements of its language-game. It seems more practical to concentrate on how to move this or that piece in the game, this or that 'chair' out of the way. Above all, the dance must go on!

Bailando resisted the pragmatic urgency of these responses, as they remained tied down, in the usual way. The urgency was understandable, since the crisis was more like war than a performance. Still, it was necessary to pause and emerge from the trenches into No Ma's Land in order to consider what was wrong, fundamentally: "Wounded Nature would then suddenly appear, would it also do so with the force of horror?" [6]. The problem was one of fundamental values: the voluntary and avoidable provocation of extinction raised ethical issues that scientific data did not answer. The problems were, therefore, philosophical and required the uncovering of the grammar of the representation of life.

Anticipating the stuck-fly metaphor he later uses, Wittgenstein says, "A philosophical problem has the form: 'I don't know my way about'" [5, §123]. One does not see the boundaries until one smacks up against them, like trying to get around Café Müller blindfolded with inadequate guidance. In the attempt to learn from Wittgenstein, we were ourselves doubly challenged: first by the philosophical issues themselves and also by groping around the 'chairs' in the way of the ideas of one of the most original, cryptic, and controversial philosophers in the history of the discipline.

We were sure of one thing, though: environmentalism counts on a variety of language-games, distinct in structure and purpose, and in the way they support and justify various ways of solving the threat of extinctions. Yet, even with this progress we had not arrived at the necessary place. In our struggle to break free of what tied us down, we found support in the philosophy of Michael Thompson and Philippa Foot (Chaps. 10–12).

The late Philippa Foot was a famous moral philosopher, the one who, among other things, invented the so-called Trolley problem [9, 10]. A run-away trolley is heading toward a group of five innocent people stuck on the track. They will all die if it hits them. It can be diverted to another track, where it will kill only one innocent person stuck there. The case has been widely discussed for the many challenging ethical issues it raises, and it has significance for practitioners of conservation because it illustrates the irreconcilability of two main approaches: consequentialist and non-consequentialist. Consequentialists maintain that we must aim for the best results, overall, which in this case seems to mean bringing about the least number of deaths. So, all things equal, the ethical thing is to redirect the trolley over the one. Of course, non-consequentialists also value good results, but believe there are ethical limits to what we can do to bring them about; we cannot, for example, actively and intentionally direct a lethal object at someone just to bring about better consequences in terms of lives lost. Otherwise, as in another example of Foot's, a hospital low on organ donations might justify killing an unsuspecting patient there, in order to harvest his organs for five others who will die without them [9]. Non-consequentialists insist on principles and boundaries that must be respected even if that involves a sacrifice in overall outcomes. And this complicates the response to the Trolley problem.

Environmentalists, in our experience, are by and large *consequentialists*, taking its pragmatic approach more or less for granted, as the paragraph cited earlier (Chap. 1) from the Brundtland Report indicates:

> No one cares for the prospect of consigning threatened species to oblivion. But insofar as choices are already being made, unwittingly, they should be made with selective discretion that takes into account the impact of the extinction of a species upon the biosphere or on the integrity of a given ecosystem [11, Ch. 6, §67].

Yet, not all practitioners believe they can do their work by fulfilling the expectations of the Brundtland Report, somehow deciding which species to let go in the name of human 'needs.' Foot comes here to the rescue. As we interpret Foot's theory, it provides foundations that can support an alternative, non-consequentialist approach, where what must be respected above all else is a creature's need to fulfill the requirements of its form, and not any overall balancing of costs and benefits. '*Natural goodness*,' as Foot defines it in her book by the same name [12], is the satisfaction of those needs, which are themselves described by the shape of the creature's life: its developmental cycle and the characteristic qualities and behaviors of its kind, those it needs to flourish in its natural habitat. In short, natural goodness consists in a creature's satisfying, autonomously, the necessities of its kind or form of life. Each form of life is autonomous in the sense that its needs are independent of the needs, interests, or desires of any other forms.

The grammar of the language used to express natural goodness—'the language of life forms,' as we will call it—has been systematically investigated and mapped out by another philosopher, Michael Thompson [13–15]. Both Foot and Thompson adhere to Wittgenstein's view that the essential task of philosophy is one of uncovering and consulting the grammar of our language, in order to clarify and treat philosophical issues that arise from its use. In this case, Foot develops a theory, based on Thompson's work, in which primary goodness lies in the fulfillment of the necessities of a form of life, i.e., in natural goodness. On this view, morality is just a sub-category of primary goodness, namely, primary goodness in human life (Chap. 19).

In his theory of the logic and grammar of the representation of life, Thompson refers us to a language familiar to conservationists, namely the language of natural history, like that typically found in nature documentaries (e.g., those narrated by Jacques Cousteau), or more precisely in authoritative monographs of the natural history of a life form. What Thompson's investigations show is that we can think of natural history as a language-game whose moves describe what a living thing *is, has*, and *does*, where the description gives specific shape to the general notion of a form of life, this form (e.g., the oak) rather than that form (e.g., the butterfly). The concept of a human being, as the concept of a specific form of life, emerges from the language which describes what a human being is, has, and does, according to its form. As we have noted (Box 1.1), the language which expresses the concept of a form of life also necessarily implies evaluative judgments about individuals bearing the form. So, for example, it is good for an adult human being that she be capable of using language, since that is a necessity of the human form of life. In related applications, the same concept makes it possible to say what it is for a life to be autonomous and flourishing. These kinds of possibilities are expressed when describing, for example, echolocation in bats or dolphins, or of photosynthesis in innumerable plants. The use of language, the ability to orient oneself in space through echoes, or of synthesizing essential molecules from the energy provided by the sun, is where we find natural goodness: the primary and autonomous good (Chap. 8).

So, the language of natural history describes what a living thing "is," "has," and "does" in the sense of those terms that give specific shape to the general idea of a form of life. And the logic of this language necessarily implies a system of evaluative judgments about what is good or bad for individual bearers of the form, independently of the interests or needs of any other life form. Foot's theory makes this standard—the standard implicit in the language that represents life—the basis of her view of moral goodness as an instance of the natural goodness of human beings. Humans walk erect, have a heart with four chambers, and also make and keep promises; the latter plays as necessary a role in a human's life as the other two. Foot finds, in the language of natural history, a *unified* grammar underlying the primary notion of goodness for all living things.

It was on the basis of these theoretical developments that we began to promote a new practice of conservation, one which might be able to finally express what is really wrong, what is lost, when we unnecessarily drive a species to extinction. We have placed our confidence in the language which represents the life forms themselves and not in extraneous standards of evaluation.

The Stage of Initiating a New Language-Game

It had become clear to us that, of all the different discourses in conservation, the language of natural history was the one that led to genuine understanding of the needs of a living thing, as well as being the one that identified the creature's satisfaction of its needs as the primary good. Therefore, any practice of the conservation of the forms of life should be guided by the standard implicit in this language and committed to the satisfaction of the necessities it describes, without compromise for the sake of some balance of unrelated costs and benefits. The practice would keep conservation consistent with the language that represents and evaluates life according to the standards of its forms. It would find in the facts of the natural history of a species the primary and sufficient reasons for making judgments about how we should relate to it, what it is permissible and impermissible to do, when what we do affects it. The reasoning would not be consequentialist and it would not let judgment about what is good or bad for another life form depend on the interests or needs of our life form.

The irony is that although conservation has always had the familiar language of natural history close to hand, it has tried to express its value judgments, and justify its interventions, in terms that represent questionable human values, values distant from what is required for other kinds of living things to live. We came to see the urgency in proposing an alternative conservation practice, one based on human natural goodness and the natural goodness of the other forms of life, setting aside all other values as secondary to these.

If the same logical grammar underlies the representation of all life—and holds the standard for evaluating what is good or bad for all the various kinds of creature—then the reasons for protecting living things must be basically the same, at a general level, for each creature of the various forms. There is a language that, as it were, protects against stumbling over chairs and that gives the user control of the rope. It is impossible to cut loose from all ties; natural historical language also has its own grammar, after all. But we *want* to be tied to its rope because if we know why someone's dance moves are being impeded, then (mixing metaphors) we know where to go from No Man's Land; if we are controlled by the right grammar, then we know how to move the rope wherever we need to go.

What is implied by all this? Where do we expect to arrive with the search light on language? Wittgenstein says:

> Philosophy must not interfere in any way with the actual use of language, so it can in the end only describe it. For it cannot justify it either. It leaves everything as it is [5, §124].

The conservation of the forms of life does not discover an unexplored language, it does not invent one, and it does not create new glossaries; it just speaks of living things, and values them, according to what they are, have, and do, in a language we all attain competence in at a fairly early age. Our task is to make this language central to the practice of conservation, to the reasons for practicing it, and the specific courses of action motivated and justified by those reasons. If this language makes an impact on our values and leads to effective solutions to everything that is wrong with the extinctions we threaten, everything will have been left as it was, in the sense of our having abandoned a misguided path that we should not have ever been on. We maintain that human natural goodness is rooted in respectful consideration of the natural goodness of the other life forms in all their magnificent diversity. It is a *necessity* of our form of life to comprehend this and to commit ourselves to exploring and applying its implications.

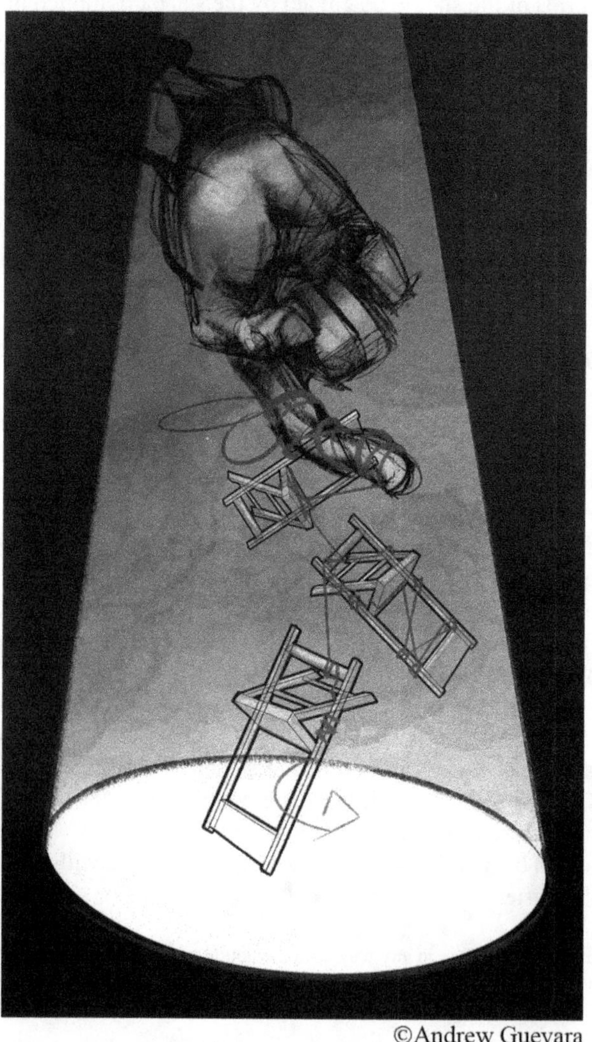

References

1. Noss, R.F., A.P. Dobson, R. Baldwin, P. Beier, C.R. Davis, D.A. Dellasala, J. Francis, H. Locke, K. Nowak, R. Lopez, and C. Reining. 2012. Bolder thinking for conservation. *Conservation Biology* 26 (1): 1–4.
2. Johns, D. 2013. The War on Nature—Turning the Tide?: Lessons from Other Movements and Conservation History. In *Ignoring Nature No More: The Case for Compassionate Conservation*, ed. M. Bekoff, 237–256. Chicago: University of Chicago Press.
3. ———. 2010. Adapting human societies to conservation. *Conservation Biology* 24 (3): 641–643.
4. https://www.pinabausch.org/work/cafe. Accessed 11 Mar 2023.
5. Wittgenstein, L. 2009. *Philosophical Investigations* (trans: Anscombe, G.E.M., Hacker, P.M.S., and Schulte, J. Revised fourth edition by Hacker P.M.S., and Schulte J.). West Sussex: Wiley-Blackwell.
6. Campagna, C. 2013. *Bailando en Tierra de Nadie: Hacia un Nuevo Discurso del Ambientalismo*. Buenos Aires: Del Nuevo Extremo.
7. Leopold, A. 1949. *A Sand County Almanac and Sketches Here and There*. New York: Oxford University Press.
8. https://www.youtube.com/watch?v=mfAN4x8MN8w&ab_channel=WatussiaMarcu. Accessed 11 Mar 2023.
9. Foot, P. 1978. Chapter 2: The Problem of Abortion and the Doctrine of Double Effect. In *Virtues and Vices*. Oxford: Basil Blackwell.
10. Cathcart, T. 2013. *The Trolley Problem or Would You Throw the Fat Guy Off the Bridge*. New York: Workman.
11. Report of the World Commission on Environment and Development: *Our Common Future* (Brundtland Report). 1967. Transmitted to the General Assembly as an Annex to document A/42/427—Development and International Co-operation: Environment. http://www.ask-force.org/web/Sustainability/Brundtland-Our-Common-Future-1987-2008.pdf. Accessed 11 Mar 2023.
12. Foot, P. 2001. *Natural Goodness*. Oxford: Clarendon Press.
13. Thompson, M. 1998. The Representation of Life. In *Reasons and Virtues: Philippa Foot and Moral Theory*, ed. R. Hursthouse, G. Lawrence, and W. Quinn. Oxford: Clarendon.
14. ———. 2004. Apprehending human form. *Royal Institute of Philosophy Supplement* 54: 47–74.
15. ———. 2008. *Life and Action: Elementary Structures of Practice and Practical Thought*. Chapters 1–4. Cambridge: Harvard University Press.

3 The Error of Wuhan

By early 2020, when we began writing this book, and at the start of the COVID-19 pandemic, it was thought that the 'wet markets' of Wuhan, China, were the likely source of the first infections—a marketplace where wild animals were bought and sold for human consumption. A respiratory virus in some of these species likely found its way to human hosts. In any case, trafficking in and consuming wild animals comes with the risk of transferring illnesses across species to human beings because it means capturing and extracting creatures from their natural habitats and transporting them for, and exposing them to, the vagaries of commerce. Good treatment of these animals—concern with their being able to satisfy the necessities of their forms of life—is not a priority, not an essential consideration, in wet markets. Therefore, many get sick and die.

'The error of Wuhan,' as we are calling it, is offered here as a poignant example of the more general error of judging as good and permissible that which should have been judged bad and unacceptable. In the markets of Wuhan, this error was, undoubtedly, committed with some frequency. The COVID-19 pandemic seems to have been one result. Then, the error is one of certain kinds of value judgment, routinely taken for granted. A "value judgment," as we are mainly using the term, is a type of judgment which—at its most general level—gets expressed in terms of what is "good" or "bad," "right" or "wrong," "permissible" or "unacceptable," etc. The more specific meaning of these expressions depends on the standards they follow in all the various contexts of their use. Without a clear standard, the use is at best incomplete, the sense diffuse. As we have also noted, value judgments support reasons for action, for eliminating a threat of extinction, for example. But if the value judgments are grounded on standards inappropriate for evalutation of the action, then the reasons for action do not justify or guide the action. If, to use a different example, we evaluate climate change according to economic standards alone, ignoring environmental considerations, it may be 'good' to continue to grow the economy that has contributed to the problem, so long as *the growth* is 'sustainable.' Reducing the concentration of CO_2 might be represented here as permissible

C. Campagna, D. Guevara, *Speaking of Forms of Life*, Fascinating Life Sciences,
https://doi.org/10.1007/978-3-031-34534-0_3

but not necessary. As a result, the grammar of evaluation impedes a movement in judgment toward uncompromising elimination of the causes of climate change.

If our primary standard had been respect for the satisfaction of the needs of a creature's autonomous form of life, then trafficking in fauna would have been unacceptable. When there is no concern for those needs, we permit actions that predispose the world to a pandemic. This was not new for the experts; the transmission of the disease was hardly a surprise. We have, then, three major crises—a pandemic, climate change, and unprecedented rates of extinction—facilitated by familiar but highly questionable values, driven by a language that takes the values mostly for granted, and that does not represent what is good or bad for living things according to their forms of life, including our own.

Environmental Origins of the Pandemic

Never before has an environmental threat impacted, almost simultaneously, so many billions of human beings. No pandemic in history compares in the potential it held to infect such an extensive population. The flu epidemic, toward the end of the First World War, was significant, but COVID-19 could have infected, and may have actually infected, a portion of the population equivalent to the entire human population in 1917–18. The world population in 1918 was 1.8 billion people, while it is around eight billion today.

None of the episodes that caused mass human deaths—like the wars of the last century, marked by genocide and violent revolutions—originated primarily in ecological disruptions. For their part, plagues, cyclones, droughts, and other calamities that could be called 'ecological' never impact all humanity and all at once. Nor do earthquakes, tsunamis, or volcanic eruptions have the potential to impact so many at the same time. They are, to be sure, events that can uproot societies, but human beings are not responsible for them. Climate change is a global, human-caused environmental problem that rivals and potentially will exceed the effects of the pandemic, but, for now, its catastrophic effects are apparently confined to particular places; they are not simultaneously affecting innumerable species. In a word, there has never been any environmental crisis like the one probably caused by us through the pandemic: a truly global environmental crisis.

It will be thought to give an excessive role to evaluative language if we propose that the human-caused, environmental event with the greatest impact in a long time was facilitated by the logic of misguided evaluative discourse about what is or is not permissible in our relationship with the other species in a wet market. However, one should consider that, whatever the final analysis shows, whether it be wet markets or something else, the source of community transmission of the virus could have been a fur farm anywhere in the world, for example, without our conclusion about the failure of evaluative language being affected. In that case, it would not have been what we are calling the error of *Wuhan*, but, nonetheless, the same kind of error. Likewise, the particular animal actually involved is not important, nor the original hosts, the intermediaries, or particular circumstances. Not even the particular virus itself is important because the same could be said of Ebola or HIV, where there was also a certain connection between human beings and other forms of life, including the underlying language contributing to the same fundamental error in judgment.

The point is that the COVID-19 pandemic is not explained by chance circumstances or ignorance. A meteorite that collides and extinguishes life is a *force majeure*; a pandemic caused by a primarily respiratory virus transmitted by trafficking in wild species is the sort of error announced at the start of this chapter. The unacceptable risk was known, it was anticipated, and it was also known how to avoid it. We understood the vulnerable contexts that could set the terrible process in motion, including the potential effects of wet markets. The likely causes were not chance or ignorance, but irresponsible decisions, which, even if not in the end the causes in this case, could easily be in the next.

In sum, in Wuhan, there were reportedly several mistakes based on poor judgement that led to problematic behavior. There was no regard for the life-sustaining needs of the trafficked animals, or the environmental consequences of such an attitude. What should have been discouraged was allowed and facilitated. It is no different with forced climate change or provoked extinction.

Return to a Familiar Language-Game

Whatever we name the error, it involves having deliberately ignored life form needs, nonhuman and human. We took wild species from their natural environments, transported them, bought, and sold them. In a market in which wildlife is traded like that, there is obviously no regard for the value of life as judged according to the satisfaction of a creature's needs. The natural goodness of the wild animal is disregarded for the sake of other values more akin to human interests and preferences, but not true human needs. Of course, our modern way of life does not see Wuhan as an error, much less as a fundamental disregard for a primary value. It is at most 'bad luck,' part of the 'cost of doing business,' or a matter of 'survival' for people with no other means of support. But what we are calling 'the error of Wuhan' has been repeated countless times and in countless places, and with many species; it is a routine of human life. There is only one language that can evaluate this state of affairs as impermissible on the standard of natural goodness, and could have perhaps avoided the consequences, by avoiding all the misrepresentation and bad judgment arising from inadequate language. The proper language is the one that represents living beings according to their natural history and by what they are, have, or do, in terms of the necessities of their forms of life.

Consider the following natural historical judgment, one of many of the sort we will be investigating: "The bat is a mammal with large ears that orients itself in space and locates food from the echo of the sounds it emits." Propositions such as these, as we have seen, and will continue to develop, have a particular logical structure that describes the living being, differentiates it from what does not belong to the category of life, and allows judgments or evaluations of the type: "This is a good representative of a bat." If a bat regularly bumps into objects in its path, then the basis for asserting that it is a poor representative (perhaps its sonar is failing) lies in the same kind of description, i.e., a description of what an individual of this form of life must be like. Rocks, rivers, the atmosphere, the planets, robots, and all inanimate things are described in a categorically different way, one whose underlying grammar does not make the same logical demands on us in our representations of them. In fact, this is the only way to know that we are dealing with non-living rather than living things: i.e., through the logic of the grammar we must use to represent them.

If based on the language proper to the representation of life, what would have been the decisions in Wuhan that might have prevented the disease's leap from wild animal to human being? We know that consideration of the natural needs of living things conflicts with wildlife trafficking, wet markets, wildlife consumption. Respect for the needs of living things as the primary value includes respect for human needs too, but we would have had to find alternatives to these markets as ways of satisfying our own needs. For it is impossible to traffic in wildlife as in Wuhan, or condone it, if one has regard for the natural goodness of living things. We can arrive at the Wuhan markets only by ignoring the standard implicit in the language of the natural history of life forms.

As we have seen, and will continue to discuss, there are many ways of representing and evaluating living things other than with the language of natural history. The choreographies of the Tokyo Olympics and Café Müller base the value of their representation on simulating what would be defective if it were ordinary human movement; for ordinarily something wrong is happening if a human being moves like a robot or stumbles into objects. The primary standard for evaluating these choreographies is aesthetic; we miss the good in them unless we apply the relevant standard. And the language we must use to describe the artistic events is not that of natural history (although the aesthetic language might fit into a larger account that describes the role of beauty and art in the natural history of human beings). Similarly, a bat can be represented as an animal that repulses some people and fascinates others. And it can also be spoken of in the language of use: "Bats serve to eliminate insects or pollinate plants for us." Neither represents the bat according to the standard of its life form, in contrast to descriptions of its echolocation. Aesthetic language or the language of use expresses value judgments based on standards having to do with human tastes or interests. This may be obvious, but then why do the prevailing practices of conservation compromise the value which would seem to be primary if the aim is to protect the various kinds of living things themselves and to promote their autonomous flourishing?

We will be looking into the answer to this question; in this chapter here we have been considering some consequences of neglecting natural goodness as the primary value. The kind of error we find in Wuhan is due to our routine acceptance of what would be unacceptable if evaluated according to the language of life forms, including the human life form. For that language forces us to represent all life forms according to the same logical grammar: bats, human beings, and all kinds of living things. This generalization is supported, as we have briefly noted, and will expand upon (Chaps. 12 and 13), in Foot's philosophical work. About us, the human being, it can be asserted—as part of our natural history—that we understand the needs of other forms of life. But if we talk of what serves us, or ask "what is it good for?" "Is it a pest?" "Can it harm me?" the standard could be purely instrumental, where what matters is only whether the life form serves as, or interferes with, a means to our ends. How we treat the life form might then be reduced to its market value, which can easily justify what would be forbidden by respect for the standard of natural goodness.

In a wet market, the relationship with wild animals is presumably strictly instrumental-economic. A language that describes how the objects of transaction live is unlikely to appear there. It is also unlikely that a caged, dead, or cut-up animal would be represented through the language of aesthetics, spirituality, or intrinsic value. Market sellers are not usually interpreters of the natural history of the species they trade, or concerned with aesthetic or spiritual values relevant to our form of life. The talk would have been of weight, parts, meat qualities, costs, food preparation, consumption, and merchandise. And even before the marketing stage, the language of the contexts of capture, transport, trafficking would have prevailed. This language carries its own standard for judgments of what is acceptable, permissible, desirable, etc., and it has a 'logical inertia' that crowds out critical or disapproving grammars. If the 'good' of an animal is that it is tender or fresh, descriptions of its social behavior find no place, the account of its natural goodness is absent, the value is instrumental.

In other words, a wet market is the place where the language on which Foot bases her theory of natural goodness is not to be found. On that theory, the primary good of a living being is to live as its nature dictates. Accordingly, whether the creature 'serves' or is 'useless' to the human, or whether it is repugnant, harmful, beautiful, or holy, is secondary, at best. Natural goodness is to live autonomously, in the satisfaction of the necessities of one's form of life. When its value is reduced to market price, the creature's form of life disappears from the language.

Wet markets are here just used to illustrate the workings of language. The errors of Wuhan are driven by language, inasmuch as language is the vehicle of thought, and they are serious mistakes because they involve how we regard everything that lives. The language ignores or crowds out the alternatives that might have averted great harm, and instead makes it seem inevitable, or routine and ordinary, whether that be the harm in forced extinctions, forced climate, or the worldwide environmental harm of a pandemic. If ethical, aesthetic, religious, or ecological standards had been prioritized instead of standards of use, different evaluations would have been reached. There are conservation practices that are grounded in the language of rights, welfare, or the intrinsic value of living things. These practices prioritize standards that allow us to make judgments of abuse, mistreatment, or immoral treatment. But 'intrinsic value,' 'the right to life,' 'moral obligation'—evaluated in accord with the cognitive capacities of a species, or its pain sensitivity—are terms embedded in language-games other than that of natural history. They all diverge and disassociate themselves from the standard of natural goodness, rather than subordinate themselves to it. As long as the language of natural goodness is not prioritized, the language of use persists; its standards remain unchallenged and continue to dominate evaluative contexts. Change requires a radical alternative at the level of the logical structure of representation, with a power equal to that which allowed the language of use to dominate.

Importance for Conservation

A main contention of this essay is that provoked extinctions result from innumerable actions routinely judged permissible but that are in fact unjustifiable causes of natural badness. And even when such actions are judged bad, by other standards, they are not opposed or corrected for the sake of natural goodness or badness. Conservation practices do not typically operate with the necessary objective of grounding their interventions in the correct reasons and standards. One can criminalize wildlife trafficking, which involves outlawing what for many language-games is considered bad or impermissible. But for a practice of life form conservation, it is the evaluation of the inhumane—human natural badness—rather than the illegal that is important. The evaluative standards of the language of 'animal rights' or animal welfare are extensions of standards applicable to humans. Their application is unlikely to generate consensus since they mix language-games across life forms. We cannot be sure that the standards grounded in the language of natural historical judgments will receive consensus, but we present many arguments showing why this language provides the best grounds for consensus.

The irony and tragedy is that a language that does not compromise the value of living things, but instead necessarily links it to the meeting of the requirements of their form, is near at hand, and has always been available, but it has been dismissed in countless decisions, guided by language structurally unsuited to represent natural goodness or badness. The needs of living things have thus been devalued and gravely neglected, making them vulnerable to disease and death and extinction. The representation of wildlife in every aspect of its treatment within the framework of use—from capture and trafficking to sale and consumption—deprived an enormity of individuals of their natural goodness, representing them as transactional objects and interfering with many other values as well, aesthetic and otherwise. It was this way by decision, by the will of human beings who understand the grammar of the representation of the living, but ignored it. The fact, among many others, that we have brought at least 9000 species unnecessarily to the brink of extinction shows there is something wrong with us and that we cannot simply dismiss, as unrealistic and impractical, the primacy of the language of life forms and natual goondess, and the need for radical change on their basis. The realistic and practical attitudes have left much to be desired.

On Virtues Associated with Redressing the Threat of Provoked Extinction

Avoiding mistakes like those of Wuhan requires a change of priority, one that gives primacy of place in conservation to the grammar of the representation of life. For this to happen, ethical virtues that move and guide the will in the right direction are necessary. Ethical virtues, we shall see, are part of the natural goodness of human beings. In the context of language, they could be understood as a disposition to use certain evaluative grammatical structures, and to reliably think and act according to, and for the sake of, the values they imply. In a wet market, virtues that might lead to prioritizing natural goodness, such as empathy, compassion, and respect for life, are unlikely to predominate. Humanity continually risks or suffers the consequences. The same is true of provoked extinction, which also results from a language alien to natural goodness. And, in this case, it is all living nature that suffers the consequences.

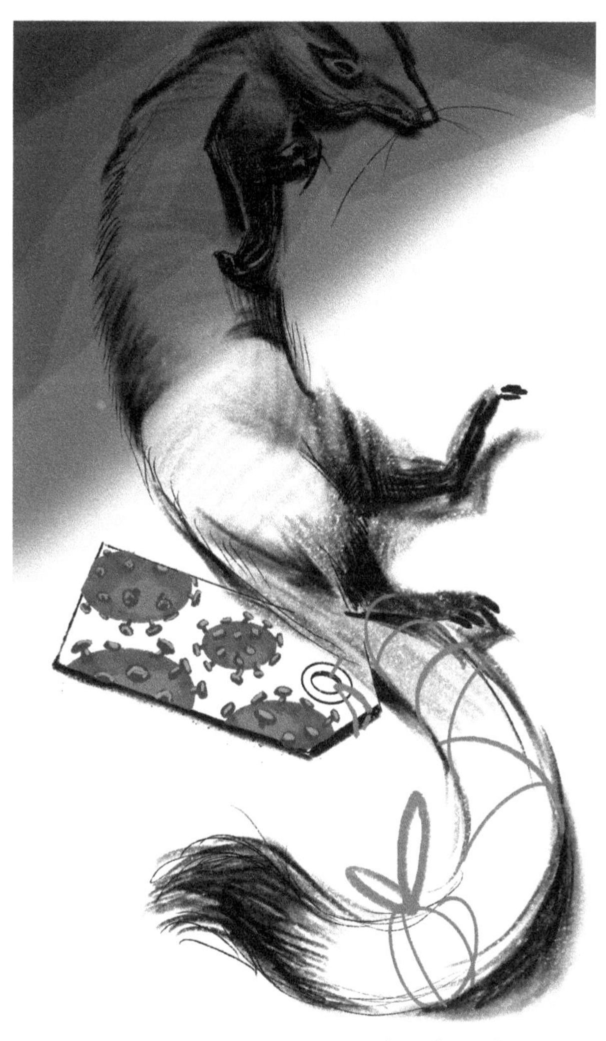

©Andrew Guevara

4 Provoking Extinctions

Like Thompson and Foot, our overall purpose in this book is 'practical-philosophical.' Like them, our theoretical investigations into the logical grammar of language have a practical purpose. In our case, we examine the language of conservation, and of provoked extinction in particular, because we seek a complete reorientation in the practice of conservation of species. We wish to ground all practice—all interventions and the reasons for them—in the proper representation of life, one that protects living things and the forms they bear, according to the standard implicit in the form-bearer relation. This is the relation described by the logic and grammar of what we are calling the "language of life forms."

We have been indicating how 'grammar' in this sense is no superficial matter of 'mere words' or 'semantics,' but rather a matter of the logical constraints on our judgments. We have proposed that the language used to represent living things in a wild-animal market made it possible to represent as good what in the language of life forms would have been categorically bad, on the standard implicit in its grammar. On the former representation, it is no surprise when seriously harmful consequences result, inasmuch as our behavior is guided by an inappropriate standard, taken for granted and distorting our view. The example of Wuhan illustrates a common and more general phenomenon of how language fosters provoked extinction. The logic of the language we use to represent living things is crucial to the value system that guides the human relationship to other forms of life. The logic can lead to errors in judgment and impede or support their correction: errors on the magnitude of the threat of unnecessary extinction, as happens when we make no distinction between natural extinction and our routine and avoidable mass annihilation of species.

C. Campagna, D. Guevara, *Speaking of Forms of Life*, Fascinating Life Sciences,
https://doi.org/10.1007/978-3-031-34534-0_4

The Passenger Pigeon

The extinction of the passenger pigeon is a paradigmatic example of voluntary human annihilation of a species. The species was typically represented as harmful, which is, most likely, one of the main reasons why it was hunted and killed to extinction. This extinction, precisely because of the underlying causes, must be represented by language other than that of evolutionary and ecological disciplines—including population statistics—where the decisions and behaviors that provoked the extinction hardly matter. The judgments that led us to annihilate the passenger pigeon were not justifiable or based on reality. The sheer abundance of pigeons was broadly believed to reassure the flourishing of the form of life against all odds, including the large-scale destruction of their breeding habitats. But animals were hunted for a variety of reasons, from the imposed costs on some who saw it as a pestilence, to the entertainment of hunting, and an over-exploitation of its population comparable to what happens today in industrial fisheries, which target many 'commercial' fishes and some invertebrates. Yet, when the killing had led to a significant drop in abundance, there was no change in the assessment of the threat of this 'undesirable' species. The change in attitude that would have avoided extinction for the sake of the life form itself would have required a change in the logical architecture of the representation of it. But the needed change in grammar did not occur. No doubt it would have involved overcoming many cognitive and emotional resistances to install a language of representation and evaluation incompatible with the familiar, undiscriminating grammar of extinction, and to replace it with the grammar of the natural history of the threatened species.

The passenger pigeon was described as migrating and breeding in large numbers, and as traveling in flocks more than a kilometer wide, which could pass for 16 h uninterrupted. That is the language of life forms. But the natural spectacle of migrating flocks was also described as a 'feathered tempest,' expressing one attitude behind our destruction of it. How was it possible to extinguish such a successful species? It was done by hunting, trapping, and poisoning them at every opportunity. Their nests and eggs were also destroyed [1]. Their migrating and breeding exposed them to hunters. The natural good of each representative depended on satisfying the needs of its form, which of course included these migrations. Paradoxically, the human beings who witnessed first-hand the satisfaction of those spectacular necessities evaluated the phenomenon as 'bad,' 'a threat,' 'a nuisance.' The dominance of one grammar of 'good' over the other facilitated the extinction.

In the extinction of the passenger pigeon, human judgment was blind to the life form's primary value, i.e., to the natural goodness of each of its living representatives, none of which will ever be seen again. Today, the shifting-baseline syndrome has [2] yielded a new 'normal,' making it impossible to even imagine the loss that we are talking about. The language that blinds us to, and perpetuates, the error of ignoring or discounting primary value for life forms does so because of its failure to represent life forms in terms of the natural goodness of their bearers.

Instead, species are treated as abstract collections, as happens in talk of 'populations.' A different outcome would have been possible through a transformation in judgment, at the level of the logic of life forms, which could properly represent the value of these great migrations, among other characteristics of the species' natural history. If the change had been achieved, it would have restored evaluative priority to the language of life forms. We maintain that a transformation in judgment is still needed today; the threats of extinction that routinely affect innumerable species will persist if we do not succeed in installing the necessary language of representation. It is worth noting that most endangered species do not have the kind of impact on human beings that the passenger pigeon is thought to have had and yet the crisis of provoked extinction is as bad as ever. The crisis is a crisis of the language we use to represent it and it will continue and deepen until we understand how that is so.

There are many conservation practices today that attempt to address the threat of human-caused extinction, recognizing provoked extinctions as bad, even, as a moral evil (as we have seen), but they argue with a different standard in view than the one implicit in the grammar of natural goodness. The IUCN Red List, for example, operates under a general assumption that a certain kind of scientific knowledge that informs us of the threat of extinction will tend to lead to our identifying it as bad or intolerable, and to its correction. The perspective seems reasonable and yet the List 'reddens' every year. The List's categories for identifying threats are concerned with 'abundance' and 'distribution,' and the language for assessing these depends on the language of 'populations.' This is the language of biology, but it is not one that expresses what living things do, have, or are, as forms of life. Something similar happens with another language-game of science, dedicated no longer to correcting the threat, but to remedying extinction itself, i.e., the language-game of synthetic biology, and in particular, of 'de-extinction,' which proposes to 'reverse' the irreversible loss of, for example, the passenger pigeon, through technical interventions [3, 4]. Whatever emerges will have very little to do with what that species was, did, or had as a life form.

The Will Comes into Play

We are distinguishing two modes of extinction: provoked extinction and the extinction of the paleontologists. In the extinction of the passenger pigeon—unlike, say, that of the dinosaurs—cultural forces were involved, on top of whatever natural forces also did. Our choices and actions—our food preference or our search for the medicinal power of a species, for example—shape the cultural threats that can lead to extinction. Many sharks are endangered because in a few cultures fin soup is consumed as a luxury food; rhinos are likewise threatened for the supposed medicinal power of their horns. These ideas, like many ideas that a species is harmful to us, may be unfounded, and yet be used to justify actions that foster extinction. Such ideas, aside from being highly questionable as justifications for provoked extinction, do not involve natural selection or chance. The passenger pigeons struck down in flight by shotgun blasts were meeting the needs of their way of life quite well. Their death was not natural.

Most of the cataclysms that affected large-scale life forms occurred in the absence of humans. In contrast, humanity is today the exemplar of these destructive forces. Human beings harm other species and their environment through what must be called a kind of super-predation, orders of magnitude greater than any made necessary by our life form, and, likewise, through a super-competition for resources, or super-consumption, of a growing super-population. These are the dominant causes of extinction. They routinely depend on decisions made in the knowledge of the consequences or willful ignorance of them. In these obvious ways, provoked extinction is a different phenomenon from the extinction of paleontologists. Why then does the dominant language of representing them homogenize the two, unifying them as if they had everything in common?

The First of Its Kind in the History of Extinction

We are concerned with provoked extinction, the loss of forms of life that could be prevented, in whose wake we find a language of conservation that reflects the decision to prioritize cultural standards of particular interest groups, over the natural good of other living beings, and not necessarily in line with the natural human good either. And although the difference between the two types of extinction—provoked by us and caused by non-human forces—is well known, there is nevertheless now common talk of the 'sixth mass extinction of species' [5]. The provoked extinction is thus integrated, by language, into a framework of events to which it does not belong. Indiscriminate language homogenizes mass extinctions, even though, as noted, it is the first time in the history of life that the loss of a life form can be evaluated as something avoidably caused by another life form and about which it makes sense to raise ethical questions of right or wrong, good or bad.

Consider the vaquita (*Phocoena sinus*), the dolphin in the Gulf of California, about to become extinct. The cause of extinction is the incidental capture of animals in fishing nets intended to catch totoaba, a fish that no one catches to eat [6]. The totoaba's value lies in its swim bladder, which is prized in traditional Chinese medicine. The fishermen who catch totoaba know that the dolphin is going extinct and that the fishery is illegal, but they continue. The catch is bycatch, but the fishing that makes it possible is not accidental. When the dolphin becomes extinct, it is going to be part of the list of species of 'the sixth mass extinction.' The IUCN will place it on the Red List in the 'Extinct' category, like hundreds of other species, many of which are only recognized by experts. Is this representation of the Gulf of California dolphin's extinction appropriate? It is likely that contemporaries of the passenger pigeon's extinction included humans who understood the unnecessary harm and loss they witnessed, even in the face of the prevailing assessment of 'pest control.' Could those who killed to extinction really plead ignorance? There was always an obvious alternative, not available in the case of a meteorite. Remarkably, even today, the passenger pigeon is on the list of species that are part of the sixth mass extinction. Is there any possible justification for that kind of misrepresentation now? Is there any better example of the persistent tendency to ignore the logic of our language and the perils of doing so?

The extinction of the passenger pigeon emerges from human practices that conservationists are now willing to judge wrong, but on a variety of flexible standards allowing trade-offs with human 'interests,' or, where uncompromising, dependent on the language of 'intrinsic' value. These judged wrongs have been identified and discussed but not corrected. The guiding standards have not resulted in a general agreement in judgment that condemns what we do unconditionally, on the basis of the natural goodness of life forms themselves, ours included. Nor, therefore, have they resulted in an uncompromising, effective, and enduring effort to redress the errors. For decades, scientific data have shown that the vaquita may become extinct [7]. And the consensus is that the species will indeed probably go extinct. But the price the market is willing to pay for a totoaba swim bladder exceeds the value of avoiding the risk indicated by the scientific data and consensus. Use value dominates. How can there be any change if the language and concepts we use to represent and evaluate the forms of life facilitate our annihilating it and blind us to what must be done to prevent provoked extinction?

Of things that must be done, some are obvious and easy: our language cannot simply lump all extinctions together, those involving human will with all others. As long as the language does not differentiate provoked from natural extinction, the human behavior that causes extinction will never be properly evaluated. Moreover, as long as what is judged bad about extinction is argued on the basis of standards that are not necessarily connected to what it is for members of a species to live, then the reasons for eliminating our threat to them will have no basis in what it is objectively for them to do well and flourish. Inattention to the misrepresentations of language leads to what we routinely do to provoke extinctions; attention to language, and in particular to what we are calling the language of life forms, is necessary to avoid this and set us on the right path. Interventions start with the language that grounds the underlying values.

Box 4.1: The Predisposition to Provoke Extinction

This story belongs to the account of the expedition of Francis Drake, who, in 1578, stopped at the coast of Argentinian Patagonia to stock up on food:

> Where into we being entered, we found within a faire and large iland, where into som of our men being sent, to see what good things it would yield for our maintenance during the tyme of our abode to do our buisines. They found it a stoare-house of victualls for a king's army ; for such was the infinite store of eggs and birdes, that there was no footeing upon the ground but to tread upon the one or the other, or both, at euery stepp ; yea the birds was so thick and would not remoove, that they were enforced with cudgells and swords to kill them to make our way to goe, and, the night draweing on, the fowles increased more and more, so that there was no place for them to rest in ; nay, every third bird could not find anny roome, in so much that they sought to settle themselves upon our heads and shoulders, armes, and all parts of our body they could, in most strange manner, without any feare ; yea they were so speedy to place them- selves upon us, that one of us was glad to help another, and when no beating with poles, cudgells, swords, and daggers would keep them ofi' from our bodyes, wee were driven with oui- hands to pull them away one from another, till with pulling and killing wee fainted, and could not prevaile, but were more and more overcharged with featherd enemies, whose cryes were terrible, and their poder and shott poisoned us even unto death, if the sooner wee had not retired and given them the field for the tyme. Wee therefore takeing with us sufficient victuall for the tyme present, took fitter opertunity of tyme the next day, and at all other tymes, to take revenge upon so barbarous adversaryes, and to weaken their power. The reason of this boldenes and want of fear I gather to bee, because they never knew what a man ment before ; for no people ever frequenting those partes, but onely the giants the inhabitants, they were never beaten or disquieted to breed in them anny dislike, for the giants themselves never use boates or com upon the water, nor so much as touch water with their feet, if they can by anny meanes avoide it. [8, Footnote of pages 46–47].

The paragraph illustrates a variety of language-games of representation and evaluation. The treatment of the birds is not surprising, given the language-games. Representing birds as a 'victuals market' led to their being killed for food, not much different from the language we'd expect at a wet market, nowadays, and the language that made possible the extinction of the passenger pigeon. Representing them as monstrous and dangerous led to their being killed in self-defense, as with the many species that have been annihilated by declaring them pests, including species that are so-called 'invasives,' many of which were introduced to alien environments by humans.

Distinctive and common human attitudes predisposed the explorers to the slaughter: e.g., outrage over what, without the slaughter, would have been perceived as an aberration in the order of things, in the need for the human being to retain its dominance over other animals. The extinction of the passenger pigeon resulted from the multiplication of episodes like those described in the quote.

(continued)

Box 4.1 (continued)

The philosopher Hans Jonas is credited with having said: "Nature's bounty, exemplifying the principle of plenitude, is a cause for hope" [9, p. 5]. The abundance, the profusion of life, the richness that Nature can display satisfies the principle of plenitude: every form of life brings richness to the modalities of existence. What happened then to Drake's men, for whom Nature's prodigality was a cause for revenge?

The stage for the killing was set by an assumed hierarchical model like that of the *scala naturae*, with the human being on a higher rung than the other animals. Faced with inferior animals, it was just and necessary to restore order by interrupting their procreation, for example. The conception of a natural ladder no longer governs the sense of nature. In the Darwinian evolutionary conception, the vertical axis is not hierarchical but temporal, and all the species present are distributed on a common horizon. However, we still speak of 'superior' and 'inferior' animals, and some sentient and intelligent species are favored (Boxes 9.1 and 14.1).

The slaughter of animals in the sixteenth century as in the nineteenth century—and before and after that too, for so many other species beyond the passenger pigeon, such as whales or fur seals, and even today, of course, with so many industrial fisheries—is contemptuous and wasteful, even when done for commercial purposes or mere entertainment. These are slaughters that are not related to the satisfaction of any reasonable value in human life, but instead, simply a reflection of a misguided predisposition of human prejudice: a recurrent theme in the chapters that follow.

References

1. http://www.audubon.org/magazine/may-june-2014/why-passenger-pigeon-went-extinct. Accessed 11 Mar 2023.
2. Pauly, D. 1995. Anecdotes and the shifting baseline syndrome of fisheries. *Trends in Ecology & Evolution* 10 (10): 430.
3. https://reviverestore.org/about-the-passenger-pigeon/. Accessed 11 Mar 2023.
4. Campagna, C., D. Guevara, and B.J. Le Boeuf. 2017. De-scenting Extinction: The Promise of De-extinction may Hasten Continuing Extinctions. In *Recreating the Wild: De-extinction, Technology, and The Ethics of Conservation*, ed. G.E. Kaebnick and B. Jennings, 48–53. The Hastings Center Report. https://onlinelibrary.wiley.com/doi/full/10.1002/hast.752. Accessed 11 Mar 2023.
5. Kolbert, E. 2014. *The Sixth Extinction: An Unnatural History*. New York: Henry Holt.
6. Taylor, B.L., L. Rojas-Bracho, J. Moore, A. Jaramillo-Legorreta, J.M. Ver Hoef, G. Cardenas-Hinojosa, E. Nieto-Garcia, J. Barlow, T. Gerrodette, N. Tregenza, and L. Thomas. 2017. Extinction is imminent for Mexico's endemic porpoise unless fishery bycatch is eliminated. *Conservation Letters* 10 (5): 588–595.

7. Jaramillo-Legorreta, A., L. Rojas-Bracho, R.L. Brownell Jr., A.J. Read, R.R. Reeves, K. Ralls, and B.L. Taylor. 2007. Saving the vaquita: immediate action, not more data. *Conservation Biology* 21 (6): 1653–1655.
8. Drake, F. 1854. *The World Encompassed*, 1st edn: 1628. London: The Hakluyt Society. https://archive.org/details/worldencompassed16drak/page/n14/mode/1up?ref=ol&view=theater. Accessed 11 Mar 2023.
9. Lee, P.A. 2013. *There is a Garden in the Mind: A Memoir of Alan Chadwick and the Organic Movement in California*. Berkeley, CA: North Atlantic Books.

5 Limits of Language

Our ability to perceive quality in nature begins as in art with the pretty. It expands through successive stages of the beautiful to values as yet uncaptured by language.
A. Leopold [1, p. 96]

This chapter explores the notion of the limits of language in conservation practice. It might be that representing the full significance of the provoked extinction of a species, with a language that does not reduce it to the language of paleontology, or natural causes not related to humans, presents us with an insurmountable obstacle. There may be values in nature for which language does not (yet?) lend itself to adequately represent. As we have been saying, and will continue to argue: If the language is inadequate, if the standard for applying "good" or "bad" or other evaluative terms, is distanced from what living beings are, have, or do according to the dictates of their form, the resulting value judgments will also be inadequate, along with our reasons to act on the basis of them. The errors may even pave the way for a provoked extinction. But you do not arrive at a bad port just from errors of judgment. Finding that our standards are empty, or that we do not have the language to express certain qualities in nature, is another way of being confronted with a chasm distancing us from an understanding of life and the living, and from being able to express the loss or harm of our extinguishing them.

An important problem with getting the language of provoked extinction wrong—either because an inappropriate grammar intrudes on it or because one cannot be found that does the job of representing it well—is that the inappropriate or inadequate does not disable the use of language. The use of language does not come with alarms that warn us when nonsense is expressed, or when an error in logic is made, or when grammars are confusingly intermingled. The language of economics, used to express the costs and benefits of provoked extinction, is hardly appropriate as a representation of the effects on the life forms themselves and the natural goodness of creatures satisfying the needs of their forms. However, superficially

C. Campagna, D. Guevara, *Speaking of Forms of Life*, Fascinating Life Sciences,
https://doi.org/10.1007/978-3-031-34534-0_5

(where grammar meets the eye, as it were), it obviously does allow us to say that an extinction is a loss of opportunity for commercial use and even that the evil of extinction lies in that loss. Our contention is that this also obviously falls well short of an adequate representation of that loss because it is so distant from an adequate representation of the extinguished life forms themselves.

We have maintained that there is a language that could better express that loss and whose grammar has the potential to capture the evil that seems to many of us to be in it, strictly on the basis of what living things are, have, and do. With this language we might find our way out of the 'bottle' that traps us, by following the logic that orders values according to the natural needs of living things first, above other standards. Before considering the advantage of this language for the representation of the loss of provoked extinction, we briefly consider the general question of the limits of language and how it could be challenged by certain configurations of reality that cannot be represented, not even approximately, by the grammar available. This is what philosopher Cora Diamond has called "the difficulty of reality" [2]. Then, it might not be simply that conservationists are misrepresenting the 'essence' of nature, but rather, as Leopold observes in the header quote, that the values intuitively but inarticulately recognized there are not yet captured by any language-game. Can they be? This chapter explores the difficulties of expressing all that provoked extinctions represent. There surely seems to be something wrong with us that we routinely and unceremoniously drive species to extinction and the proposal in this essay is that the language of life forms is the only sound basis for trying to articulate this. But are we really capable of comprehending what is wrong and lost, especially in a way that can finally transform the world for the better?

The Inexpressible

Unless we are prepared to recognize limits of language when we hit them, we risk a kind of bewitchment induced by the impression that we have expressed all that needs to be said about a difficult reality. In the case of the threat of provoked extinction, this sort of bewilderment has led to attempts to redress the threat for reasons that have no necessary connection to the threatened life. We are lulled into a kind of complacency and false assurance, imagining that our words have done the job, when in fact they have fallen far from the mark.

The difficulty of reality can be illustrated by certain extreme situations, such as concentration camps, life on a trench front, or the Chernobyl disaster. Can we ever really express or comprehend the loss, the harm? Exceptional cases are not required for reality to seem inexpressible: Diamond has us consider a photograph of loved ones, in their vigorous and youthful beauty, all of whom have since died. We can describe the picture, as we just did. They are dead. They were once alive and young. What is the problem? And yet, what can be said about the state of affairs in this way does not even approximately represent our full experience; it is like geometrical language attempting to represent ordinary human motion. The one who faces the difficulty of reality senses the mismatch, and feels trapped by the language, but does not necessarily glimpse a way out, as may have been the case for Leopold in confronting the value in nature.

As we have been saying, conservation practices dealing with provoked extinction do not feel compelled to consider the limits of their language in considering causes of extinctions and their correction. There is a tacit assumption that we can represent and evaluate any state of affairs, with the language at hand. The differences between evaluations are explained by the prioritization of relative standards. Some practices prioritize use, others beauty, etc. Thinking about language, questioning it, doubting it, and being alert to its limits would seem to be the domain of disciplines that do not solve problems of species conservation; investigating limits in language is not part of environmentalist pragmatism. And yet it is Aldo Leopold, in his environmentalist masterpiece, who invites us to attend to the limits and possibilities of language and to make that essential to the art and practice conservation.

The Great Silence

If the reality of provoked extinction poses a difficulty for human comprehension, escaping full representation in our language, then we must consider one of Wittgenstein's best-known sayings: "[w]hereof one cannot speak, thereof one must be silent" [3, p. 90]. Of course, we do not see this as advice to downplay the role of language, or to adopt the more familiar and dominant language of conservation, ignoring how it may contribute to the threats to life, or to misguided interventions based on implicit, inappropriate, or inadequate standards. No. Rather, we take it as an admonition that it is better to be silent than to be satisfied using language that ignores crucial realities while pretending to speak of them. For example, we cannot be content with describing the COVID-19 pandemic with the language of public health nor the intervening to correct it with the language of medical sciences, as though that were all there is to it. We cannot ignore the fact that the event is a worldwide environmental problem. Nor can we ignore the standard of natural goodness in assessing the harm done to species involved, human beings among them. The question is whether any of this is sufficient to the task and not to pretend that it is if it is not. We must face our limits if there is any chance of getting beyond them. A multiplicity of grammars does not make up for imprecise, imperfect grammar used to describe the nature of the problem; 'pluralism' does not compensate for what is under-represented or misrepresented (Chap. 16), namely, the life form needs of living creatures.

We must overcome a certain kind of confusion Wittgenstein thought common to philosophy's characteristic inability to make progress. As he observes, "[t]he confusions which occupy us arise when language is, as it were, idling, not when it is doing work" [4, §132]. While a car is idling in neutral, the engine gets no traction; even if the car somehow moves, it is not driving. The metaphor depicts what Wittgenstein thought often happens in philosophy, where language seems to be doing real work, but in fact has gone 'on holiday' [4, §38]. Thus, in his view, it is the main purpose of philosophical practice to recognize when this happens. If there are 'results' in philosophy, they amount to "the discovery of some piece of plain nonsense and the bumps that the understanding has got by running up against the limits of language" [4, §119]. The solution is to bring words back to their everyday use, as he teaches in a passage worth quoting in full [4, §116]:

> When philosophers use a word—"knowledge", "being", "object", "I", "proposition/sentence", "name" and try to grasp the essence of the thing, one must always ask oneself: is the word ever actually used in this way in the language in which it is at home?—
>
> What we do is to bring words back from their metaphysical to their everyday use.

Conservation is a philosophy in need of such admonitions, especially concerning its use of terms like "species," "population," "extinction," "nature," and above all "life." These words can have special, technical meanings, in the service of conservation science or of one 'environmentalism' or another, but we must watch that they do not spin without traction, or worse, move in unguided or misleading directions when they are supposed to engage and transform our everyday life and understanding.

When we get clarity and understanding by bringing language back to its 'natural habitat,' as it were, there is a sense in which we have left everything as it was in the first place. We will devote Chap. 25 to this idea. Our message for conservation practice is to consider how a return to the language of life forms achieves this kind of clarity, and keeps us focused on the relevant reality, and the standard for evaluating it, and prevents the nonsense of leaving the rationale for action to a cost-benefit assessment conditional on political will or the availability of public resources, as the Brundtland Report has suggested. It is the language which must ground any attempt to confront the difficulty of the reality of routinely, willfully, and unnecessarily provoked extinction.

Given the dominant language-games of conservation, the extinction provoked should render us mute. The choreographer William Forsythe says of the one who expresses himself through dance,

> You don't want to be good, you don't want to be even great. You want to have people be at a loss for language... That's when dance is really working, when it kind of actually makes language not useful [5].

The dominant practices of conservation should recognize that life's greatness— its 'dance,' the way it really 'works'— makes their language useless. The highest priority in the conservation of species must include a search for the language that can properly address the difficult reality of the harm and loss of provoked extinctions, if it hopes to stop or prevent them. None of the dominant language-games with which provoked extinction is represented do this. Conservationist practice is based on having a 'solution' at hand, consensus (politically correct), 'objective' data, and the weighing and balancing of costs and benefits. And if in the process the species becomes extinct and the threats deepen? In such a case, there will always be some way of saying something that will appear to be reasonable; herein lies the spell of language.

Facing the Difficulties of Language and Reality

It is part of the practice of the conservation of life forms to show the inadequacy of the prevailing language of representation and evaluation of provoked extinction, to deepen our understanding of the alternative language of life forms and the system of values derived from it, to strengthen the use of that language, and to derive from it alone the reasons for doing conservation and implementing what it takes to redress the crisis. The inadequacy of certain philosophical views, according to Wittgenstein, is discovered, we saw, by a "grammatical investigation" (see citation in Introductory notes) that clears up misunderstandings arising from the use of words in different "regions of our language" [4, §90].

The regions of language that conservationists draw from to evaluate a provoked extinction, the arguments with which they try to justify their evaluations, and upon which they base their interventions, are misleading, confusing, and counterproductive. This can be shown through a grammatical investigation, which first requires a willingness to initiate it. To reach this point, it is necessary to generate in conservationist practices a cautious distrust of its language. This is a goal of our work. Extinction then needs to be represented by regions of language other than the dominant ones. We would reverse the usual approach in the practice of conservation and instead start with talk of life and natural goodness: i.e., the language based on the logic and grammar of the representation of life forms and the standard of value implicit in it, one that keeps us from the language that misleads us and that prevents us from overcoming the obstacles we continually face in realistically and effectively redressing the harm and loss in provoked extinction.

However, for all our optimism, it may also be that even with all the effort involved in maintaining attention to language, installing the right grammar, and intervening to set things right, an insurmountable obstacle may still present itself, as it does with certain inexpressible difficulties of reality. The writer Svetlana Alexievich makes the point, in speaking about the Chernobyl tragedy [6]. She writes: "[w]hat happened in Chornobyl[1] is much worse than the gulags and the Holocaust..." And she adds: "... to fully comprehend what is happening we would need different human experience and a different inner instrument, which does not exist yet." The limit could transcend language; it could be a limit in the very cognitive structure of human beings to represent certain realities, where the unspeakable stems from the inability to experience what needs to be said. We are investigating whether the language of life forms can assist in the generation of that 'internal instrument' which, in Alexievich's opinion, does not yet exist for tragedies like Chernobyl and, for all our efforts in this essay, may not exist for capturing all that is lost when we annihilate a species.

[1] The publication uses the Ukrainian spelling of the more familiar Russian alternative, "Chernobyl."

References

1. Leopold, A. 1949. *A Sand County Almanac and Sketches Here and There*. New York: Oxford University Press.
2. Diamond, C. 2003. The difficulty of reality and the difficulty of Philosophy. *Partial Answers: Journal of Literature and the History of Ideas* 1 (2): 1–26.
3. Wittgenstein, L. 1922. *Tractatus Logico-Philosophicus*. London: Kegan, P., Trench, Trubner.
4. ———. 2009. *Philosophical Investigations* (trans: Anscombe, G.E.M., Hacker, P.M.S., and Schulte, J. Revised fourth edition by Hacker P.M.S., and Schulte J.). West Sussex: Wiley-Blackwell.
5. Menegon, L. 2017. In reshaping an early masterpiece, choreographer William Forsythe keeps ballet on its toes. WBUR (Boston's NPR). http://www.wbur.org/hereandnow/2017/02/24/william-forsythe-artifact. Accessed 11 Mar 2023.
6. Radio Free Europe/Radio Liberty. 2015. Nobel Winner Alexievich: What she's written and what she's said. https://www.rferl.org/a/nobel-alexievich-belarus-what-she-wrote-said/27295565.html. Accessed 11 Mar 2023.

6 Glossaries, Euphemisms, Metaphors, Analogies, and Catchy Words

> *[O]nce a commitment to a particular representation of life is made—material, discursive and social—it assumes a kind of agency that both enables and constrains the thoughts and actions of biologists. In a sense, it is the representation itself that guides the imagination and reasoning.*
> *Lily E. Kay [1, p. xviii]*

In the header quote, historian of science Lily Kay observes that the way biology represents life affects how it tries to understand what it studies, for example, the genetic code. Biology 'takes on' DNA as a representative entity, which "enables and constrains the thoughts and actions of biologists." The point is especially relevant to species conservation, where the dominant language has showcased various representations of life, including the concept of DNA as definitive of life, but which are obstacles to a representation of the forms of life and their provoked extinction. For example, a tempting but mistaken idea is that large and complex organic molecules, with DNA as the quintessential example, could be the definitive characteristic marks or criteria of life. "Living things contain complex molecules, such as DNA" may then be thought as a way of representing life and differentiating the living from the inert or 'inanimate.' But as Thompson points out:

> The judgment about DNA [that is: "Living things contain DNA"], if it were true, would only show how resource-poor the physical world really is. It could make no contribution to the exposition of the concept of life, or to a teaching on the question, what life is... [2, pp. 36–37] (See also footnote 3, p. 56)

Full appreciation of Thompson's point requires considering the logical grammar of "life," an analysis we give in Chap. 10. But the essential insight is that molecules, like any other empirically identifiable, individual things, do not give us the concept of a living thing unless they are placed in a broader context than can be determined by their physical and chemical properties here and now; there must be reference to

C. Campagna, D. Guevara, *Speaking of Forms of Life*, Fascinating Life Sciences,
https://doi.org/10.1007/978-3-031-34534-0_6

something like the developmental concept of a life cycle, or, in a word, a form of life, which requires a different logical grammar.

This is our primary example of how language is a vehicle for thought, the way Wittgenstein and others teach (see [3], see also: Introductory notes and Chap. 16), and of why we must make a grammatical investigation part of conservation practice. Environmentalists such as Leopold, and also more recent ones (Chap. 16), have identified language as important to their practice, although the more recent ones have generally confined the issues to the use of isolated terms and phrases, such as 'ecosystem services,' which are obviously problematic. In our essay, the focus is on the logic underlying the use of these words and phrases and the grammar or rules for putting them together in meaningful ways. Words and phrases take on the meaning they have by the way they are put together and used in the various contexts of human life. Addressing the problems generated by language is not just a matter of replacing or coining an isolated word or phrase. A main purpose of this chapter is to examine this important point.

Creating Language

Until the nineteenth century, climate scholars were limited in the scientific language used to describe the phenomena of the atmosphere [4]. The course of weather had always been a mystery more easily integrated with opinions than with scientific causes and consequences. Meteorological science began to emerge as certain useful terms began to fill the gaps in language. Clouds, for example, received their current names (stratus, cumulus, etc.) in 1803, and the Beaufort scale, to indicate wind speed, dates from 1805.

Explaining causes and consequences of climate required new, more precise, terms, and getting rid of old vague ones: not at a simple matter of coining new terms, but of learning a new, scientific language-game, with great predictive and explanatory power. The same occurred in other disciplines. The sociobiological approach in the behavioral sciences, for example, was based, in part, on the language of effective metaphors, such as in *The Selfish Gene*, the popular book by Richard Dawkins. Robert Trivers—whose work is a pillar of sociobiology—justifies the use of metaphorical language this way:

> This was a use of language that social scientists and others would come to detest. When they were aroused in full antipathy toward "sociobiology," you often read that this was a perversion of language. True altruism had to refer to pure internal motivations or other-directed internal motivations without thought or concern for self. To an evolutionist, this seemed absurd. You begin with the effect of behavior on actors and recipients; you deal with the problem of internal motivation, which is a secondary problem, afterward. If you made this point to some of these naysayers, they would often argue back that, if this were the case, then Hamilton should have chosen words that had no connotations in everyday language. This was also a very short-sighted view. In the extreme case, I suppose, Hamilton could have called the behaviors x, y, w, and z, so as to avoid any but alphabetical connotations. But this, of course, meant that you would always have to be translating those symbols into some verbal system that made sense to you before you could think clearly. Incidentally, when the great sociobiology controversy did roll forth, I soon came to see that the real function of

> these counterarguments was to slow down your work and, if possible, stop it cold. If you start with motivation, you have given up the evolutionary analysis at the outset. If you are forced to use arbitrary symbols, progress will be slowed for no good reason. Even the invitation to argue with them seemed to me to benefit them by wasting your time! [5, pp. 6–7].

There is more than one relevant observation for us here about the history of this heated controversy. Trivers thought that ultimately what really put off critics of so-called sociobiology were the metaphors, like 'selfish gene.' William Hamilton, who laid the main theoretical foundations for evolutionary social theory, needed the metaphors in order to communicate profoundly novel and difficult ideas (a good example of the indispensability of metaphor, even for scientific understanding). Could the controversy have been resolved by just changing the names of the ideas, or even by just using mathematical variables? It could not, since understanding the terms involves understanding their role in the novel language-game, which metaphors help to communicate by relating the ideas to regions of language whose game we already know. Trivers was well aware of the effective use of metaphors:

> To discuss the problems that confront paired individuals ostensibly cooperating in a joint parental effort, I choose the language of strategy and decision, as if each individual contemplated in strategic terms the decisions it ought to make at each instant in order to maximize its reproductive success. This language is chosen purely for convenience to explore the adaptations one might expect natural selection to favor [5, p. 75].

As in meteorology and sociobiology, environmentalism has demonstrated its own processes of language modification, from Henry David Thoreau's poetic naturalism in *Walden*, to the technical terms of 'conservation biology' [6], 'conservation science' [7], and 'ecological economics' [8], among other language-games. All have generated new terms and phrases, many of which became quickly entrenched and iconic. The irony is that the terms have been so catchy, when in fact resistance was necessary in this case; unlike the cases where new language advanced understanding, some language-games of environmentalism have impeded it, at least in relation to provoked extinctions. Indeed, as we will soon see, some central terms themselves—such as 'environment' (and even 'nature')—are so varied in their uses that their contribution to the discourse is dubious. Others, like "biodiversity," could have been effective, when first coined, at stressing the value of variety in life, but ended up replacing "wildlife" and being co-opted by technical language that was incorporated into pro-development discourses, attempting to sound 'greener.'

Glossaries

Words and phrases adapted to environmental disciplines are numerous and always increasing: 'restoration,' 'sustainability,' 'resilience,' 'governance,' 'ecological rehabilitation,' 'ecosystem services,' 'carbon offsets,' 'adaptive management.' Among the phrases common to conservation biology, we have: 'assisted migration,' 'facilitated adaptation,' 'genetic rescue,' 'de-extinction.' The environmental disciplines have invented terms like 'defaunation,' 'biophilia,' 'rewilding.' And strange words have been proposed for common occurrences: 'theriocide,' for the act of killing animals in quantity [9]. Environmental problems even have their psychological affliction: e.g., 'solastalgia,' the anguish over the destructive transformation of the environment in which human life takes place [10].

The plethora of apparently innocent but tricky terms has not pass unnoticed. For example, environmental journalist George Monbiot has proposed abandoning words he calls "constipated terms": vague, technical, improper descriptors of environmental drama [11]. Monbiot includes the word "environment" among them, and adds a point we have been emphasizing:

> When a species is obliterated through human action, we use the term 'extinction'. This conveys no sense of agency, and mixes up eradication by people with the natural turnover of species [11].

Cognitivist scientists have shown that a word is enough to predispose us to take on a certain attitude:

> We find that even the subtlest instantiation of a metaphor (via a single word) can have a powerful influence over how people attempt to solve social problems like crime and how they gather information to make 'well-informed' decisions [12].

For example, if a criminal act is spoken of as a disease, the representation tends to provoke demands for preventive and diagnostic solutions. If it is presented as an act of predation, it generates a desire for prosecution and punishment. 'War' as a metaphor has been used in several studies to test human attitudes regarding environmentally pertinent issues, including climate change [13]. Researchers conclude that "when it comes to choosing a metaphor to talk about climate change, the war metaphor is consistently more impactful than the race metaphor." The 'war' against climate change is a call to battle; the 'race' against it induces attentiveness to efficient use of energy or accelerated efforts toward a miraculous technological fix. If the single word can, so to speak, carry enough value to generate new attitudes ([14], and references therein [15, 16]), the language-game can be expected to be pervasive.

We must investigate the language of conservation practices with an eye to such phenomena, as many already do. More important, however, is how those terms are put together within the rules of their language-game to carry the significance they do. Three ways of assembling words in a language that have received particular attention in environmentalist thinking are euphemisms, metaphors, and analogies.

Euphemisms

A euphemism substitutes a softened expression for a harder, repellent version of the same. In ecological studies, for example, it is often said that a population was 'sampled.' The term helps to legitimize the fact that animals are often being killed for the purposes of the study, which may relate to determining diet, reproductive condition, age, etc. [17]. An article in Spanish whose title may translate as "Ironic glossary of euphemisms beginning with 're'" [18] illustrates a variety of euphemisms specific to conservation ecology: 'restocking,' 'reforestation,' 'reintroduction.' The prefix "re" intensifies the act referred to, by indicating repetition of the act. The terms conceal and distort the harsh realities behind the issues and put a distance between them and the agent of change. Indeed, in a context of corporate social responsibility, it was shown that ethical actions were judged more favorably if presented in euphemistic terms, rather than in plain language [19]. As we have noted, the language of extinction is co-opted by euphemisms for the acts of voluntarily killing off an entire life form. Species 'disappear' or 'become extinct,' when they are actually actively annihilated, exterminated, razed, and destroyed.

Two of today's most misguided euphemisms are 'climate change' and 'sustainable development.' The former is used for our forced degradation of planetary life support mechanisms and the second for our exploitation of nature. An analysis of wording in the communication of climate change concluded that euphemisms "can undermine the objectives of raising climate change awareness and changing behaviors to reduce emissions [15]." Some words and phrases regarding climate are altogether avoided by governments, while stressed by the media, the former, for example, preferring "climate skeptic" and "climate change," while the latter speaks of "climate science denier" and "climate crisis" [16, 17]. The choice of terms seems critical if we hope to engage public interest in the causes and consequences of the human-induced climate disruption, and here the phrase "global warming," though far from perfect, serves better than "climate change" for public understanding [20].

'Sustainable development' has been possibly the most damaging euphemism in the context of the threat of provoked extinction. It attempts to speak, in terms which sound attractive both economically and environmentally, of a supposedly reasonable use of nature as an instrument to our ends. The phrase is synonymous with "conservation" in contexts where the instrumental value of nature dominates. We have discussed elsewhere that it functions as a failed *deus ex machina* [21]—that figure from ancient theater who was lowered onto the stage by a crane, and whose role was to resolve the tragedy with a miraculous act. The prevalence of this euphemism in conservationist discourse is strategic. In 1987, the Brundtland Report, the critical document that made famous the notion of sustainable development,[1] stated:

> Species problems tend to be perceived largely in scientific and conservationist terms rather than as a leading economic and resource concern. Thus the issue lacks political clout [23, Ch. 6, §55].

The report recognizes the importance of the language we use to represent an issue and proposes a change in language, in order to represent our use of nature as necessary to our welfare. The change was in fact brought about, in stages. First the concept of the environment was confined to the *human* environment. Eventually

[1] See [22] for a detailed historical account of the origin and use of the concept.

"environment," in this sense, was associated with the phrase "sustainable development." Finally, there was a total displacement of the word "environment." The names of the international conferences known as Earth Summits illustrate this:
1972: United Nations Conference on the Human Environment
1992: United Nations Conference on Environment and Development
2002: The World Summit on Sustainable Development
2012: United Nations Conference on Sustainable Development 'Rio+20'

Judging something 'sustainable,' in, for example, the sense of 'sustainable fishing,' is consistent with the idea that, for the sake of 'development,' it can be acceptable, or even desirable and good, to follow a practice of indefinitely killing representatives of countless life forms, so long as the possibility of continued killing does not decline, i.e., so long as it is 'sustainable.' And even when it does decline, when the target population is affected by extractive activity, the language of sustainability accommodates this and justifies it for the sake of continued development, with narratives related to the need to maintain jobs and reduce poverty, for example. Meanwhile, society's inequities are not seriously addressed nor their elimination identified as fundamental to the solution of environmental issues.

Some environmentalists, in their work for non-governmental organizations, align themselves with 'sustainable development' by adopting its language (Chap. 18). Some may also be aware of the problems with the language, but their attitude is that working from within is more effective than trying to oppose the concepts and language in principle. Of course, this aids and abets the instrumentalist, economic models of 'informing development,' as the eco-pragmatists propose (Chaps. 16 and 18). Adopting the metaphors is the first sign of camaraderie. It has been suggested that these conservation movements should be held co-responsible for the annihilation of species [24], inasmuch as they support value systems that tend to cause it.

Metaphors

Philip Ball [25] has observed that:

> In literature, metaphor serves poetic ends; in politics, it is a (subtly manipulative) argument by analogy. But in science, metaphor is widely considered an essential tool for understanding...

As we have noted, "the selfish gene" is a metaphor of widespread use in the popularization of sociobiology, according to which, among other things, the discussion of the evolution of 'altruism' is based. Conceptualizing DNA as a 'code' was essential to the design of experiments that uncovered the relationship between DNA's structure and protein synthesis [1, 26]. Representing DNA as a code made it easier to conceive of research as a decipherment. Upon understanding the code, and concomitant with the development of informatics and computer science, DNA was represented as the 'software of life,' another metaphor, and one that inspired the development of synthetic biology as a branch of engineering applied to 'creating' living things [27]. Finally, metaphors also influence our approach to the crisis of provoked extinctions, when variously represented as a "mistake" (e.g., as a loss of resources), or as a "crime" (as, for example, one of interspecies genocide), or a "cancer" and thus as a deadly illness, and (perhaps) unavoidable consequence of human behavior [28].

Analogies

Analogies, which include metaphors as a key type, are based on the similarities found among distinct things. Climate change, for example, is often treated as a phenomenon analogous to a disease, as just mentioned [29]. This use of language is meant to help us understand something problematic or difficult, by speaking of it in terms of something else that resembles it and is better understood. Confusion arises when the analogies are stretched too far across loose similarities, exploiting misleading convergences in the rules between language-games. For example, a recent newspaper article has as its title: "Mushrooms communicate with each other using up to 50 words..." [30]. Research shows that under certain conditions some species of fungi can send electrical impulses which could be interpreted as an exchange of information between individuals [31]. The analogy with human language leads to calling these impulses 'words,' suggesting that fungi have a 'language.' Another example is the application of the notion of extinction to human languages. One speaks of extinct or dead languages. Unlike the disease analogy for climate change, these two examples are not necessarily meant metaphorically. However, their literal use is a stretch, neglecting or ignoring the differences in grammar of the analogized terms and exaggerating the similarities [32].

Failures of Vigilance

Analogy, metaphor, and euphemism all have their function in language, their proper place; conservation and environmentalism must utilize at least some of them in dealing with the complex and challenging realities it wishes to comprehend and evaluate. However, through lazy metaphors, buzzwords, and politically and culturally safe rhetoric, the discourse of conservation has fallen into what, in another context, philosopher Cora Diamond, following Stanley Cavell's writings, calls 'philosophical deflection' [33]. In her own account of the concept, she illustrates it within the context of philosophical responses to J.M. Coetzee's novel, *The Lives of Animals*. The novel is concerned with issues related to our treatment of other animals, especially the horrors of factory farming and the like. The central characters are developed with a view to portraying a variety of views, but there is no sustained argument that could summarize the novel's 'point' or position, as falling on one side or the other. It is not philosophy in that sense common to professional, academic philosophy. And, as Diamond compellingly argues, the philosophers who have tried to interpret the novel with philosophical language, turning it into a philosophical argument, are deflecting from what the novel really does: namely, present us with some very difficult realities, whose difficulty is often betrayed by our finding ourselves in evident contradiction or confusion when try to speak about it and comprehend it (as the main character and others find themselves at times in the novel). Philosophy deflects because it has to clean up all such difficulties for the sake of a coherent argument and view.

A philosophical deflection tends to arise when reality overwhelms a language-game's modes and capacities for representation and rather than acknowledge the limitation, the language falls back on available moves that do what they can but without really coming to grips with the difficulty. The distraction can be effected by euphemisms, inept metaphors and analogies, and many other ways, including flatfooted literalness or 'factualness.'

The problem with deflection is that it is an avoidance and distortion of the very reality that prompts it. The metaphors and euphemisms of conservation biology, those of the eco-pragmatists, or those who sympathize with sustainable development, are avoidances of the reality of provoked extinction and its implications for the forms of life. Even the *ethical* representation of provoked extinction as a "moral wrong" [34] can be a deflection of language, inasmuch as the relationship between ethical language and the language of life forms, and their natural goodness, is not yet made clear (Chaps. 19 and 20). The reality of provoked extinction involves unnecessary, relentless acts of killing until extinction, which we can only begin to understand the significance of when we represent them, for example, in quantified-statistical data. But we must emphasize that that is only the beginning. A comparison to genocide is apt (even if the moral equivalence is questioned) because, there too, the harm or loss transcends the sheer numbers, though they surely count. At this stage of our understanding, it is an avoidance, however right it may feel, to simply use the words "moral evil" to describe the harm and loss of the extinctions we unnecessarily cause, in the same way we uncontroversially use them to judge the harm and loss of the Holocaust. We had to learn the grammar or language-game of *genocide* to really comprehend that new horrific reality, to the extent that we have comprehended it.[2] Understanding what goes into developing a new grammar, for such a disturbing reality, may help us with provoked extinction, but this is not a matter of simply applying familiar terms, which we have learned primarily in human contexts, to radically new issues raised by our effects on all the other forms of life.

All in all, language is used to avoid reality in many ways: language can make a reality that is incomprehensible to us appear as though it were perfectly clear, the simple and understandable appear complicated and unintelligible, or the nonsensical and unintelligible appear astute and enlightening. In the novel, *Being There*, novelist Jerzy Kosinski tells the story of an unknown, profoundly unimpressive and

[2] Which was not the simple coining of a new term, but the hard work of writers such as Primo Levi, Hannah Arendt, and others.

inarticulate man, who, through a comedy of accidents, winds up becoming a respected and influential voice in high political places because his simple and empty discourse is interpreted as the enigmatic metaphors of a man of great wisdom. Conservation practice is susceptible to the '*Being There* effect,' as we might call it, insofar as it has allowed words, dressed in good seeming rhetoric, to lose all recognizable connection with the reality they mean to address.

We have cited Wittgenstein's famous saying that philosophy is a struggle against the bewitchment of our understanding by means of our language [35, §109]. The bewitchments are disarmed by means of grammatical investigations and prevented by alertness to the logic of our representations. 'Ecosystem services' is a metaphoric euphemism for commodifying and instrumentalizing planetary functions, with the perhaps well-intentioned purpose of inducing us to see nature as valuable. It exploits our preoccupation with economic profit, which is indeed great in at least a very powerful subset of society. Predictably, it has displaced other ways of representing and evaluating nature, to the point of making it seem essential to understanding our relationship with nature, as Richard Norgaard points out [36]:

> What started as a humble metaphor to help us think about our relation to nature has become integral to how we are addressing the future of humanity and the course of biological evolution.

The cognitive research of Paul Thibodeau and others suggests that "[f]ar from being mere rhetorical flourishes, metaphors have profound influences on how we conceptualize and act with respect to important societal issues. . . " [12]. The key metaphors of environmentalism are among the best examples of this. If life forms are represented as service providers, they will end up being understood and treated with the language that represents telephone companies or airlines, for example. The relationship with them will then be based on trying to extract the maximum benefit at the lowest economic cost. This is the approach of the Brundtland Report. If an extinction is included in the discussion of the 'opportunity cost' for development (e.g., the loss of the opportunity for extracting some useful drug), it is also suggested that this cost can be balanced out by an alternative benefit. It has been said that "[w] ords can be like tiny doses of arsenic: they are swallowed unnoticed, appear to have no effect, and then after a little time the toxic reaction sets in after all," [37] and also that "[t]he price of metaphor is eternal vigilance."[3] Orwell's *1984* made the point famous by imagining what results when ideological metaphor and politically perverse language thrives and impedes understanding of the human form of life.

The metaphor, the euphemism, and the catchy word can install a language-game. As we have tried to emphasize, it is important to think of the word, its effect, its significance, as possible only because of its place in the broader context of the language-game. That is part of the point of the analogy to games. There is no sense in trying to understand a chess move (e.g., checkmate), by concentrating on just one

---

[3] Attributed to the cyberneticists Arturo Rosenblueth and Norbert Wiener by Lewontin, R. C. [26].

move. You have to have the broader knowledge of the game, to some extent, in order to know any part of it.

The grammar or logic of a language can be compared to 'rules' which facilitate or oppose, stimulate or discourage, allow or forbid certain 'moves' in a game. Criticism of the language-games of conservation should not then be limited to piecemeal revision of the glossary of isolated words or turns of phrase. The evaluative bias of the phrase 'natural capital' is not only explained by looking up a definition of the two words, but by the power they have in certain language-games to enable or impede certain moves, which invite one particular gloss and reject another, one funding source and not another. It is in this way that language confines or liberates, leading to constructions that represent a state of affairs from a limited or expanded menu of possible values—leaving out some values, including others—making some seem possible, and others impractical, unreasonable, unimaginable.

The instrumentalist-economic metaphor now thrives in the language typically used by conservationists to represent life. The standards of evaluation of this language-game have been in place since the strategic dominance of the Brundtland Report and reinforced by decades of funding for environmental interventions that rewards those who best adhere to the logic of development. This is why some environmental organizations have even changed their missions and visions, a transformation driven by a political and funding strategy. The euphemism of 'climate change' has been so successful that it has overshadowed, for example, the language of 'biodiversity,' already degraded by economic discourse prevalent since the report [38]. Of the vast environmentalist glossary, the phrase that dominates Google searches, for example, is precisely 'climate change.' In contrast, the use of words and phrases like 'endangered species' and 'biodiversity' has fallen precipitously over time [39]. The critical threat of extinction through our over-exploitation of nature and environmental degradation of it has lost relevance in these contexts. The ignoring of the role of our agency and the elimination of responsibility that results from amalgamating the language representing human-caused extinction with other mass extinctions, coupled with the passive metaphor of 'climate change,' has depleted conservation language from its power to represent and articulate our condemnation of provoked extinction for what it seems to many of us obviously to be, namely unacceptable.

An Escape from Bewilderment

It will not do simply to banish certain metaphors or euphemisms, if indeed they could be banished, nor simply to combat them with alternative metaphors or euphemisms. It is not possible, as some have suggested, to combat the economic language of 'capital' and 'services' for nature with a pluralism of metaphors tailored to neutralize the problem [40, 41].

There is a character in Ursula Le Guin's short story "*She Unnames Them*" [42], disenchanted with language. Disillusionment leads her to 'unname' living things, emptying the language of their names. The names of insects drift away in vast swarms of syllables that buzz, sting, or dig tunnels. The names of fish drift silently away, lost in the ocean like the ink of a cuttlefish. Once the protagonist unnames every last life form, she achieves a closeness with them that names had prevented. Given how provoked extinctions are actually represented today, an act of 'unnaming' might be a way of following Wittgenstein's suggestion about keeping silent about what cannot be spoken. Or perhaps it would be appropriate to proceed as Gabriel García Márquez recounts in the opening paragraph of *One Hundred Years of Solitude*. In the early days of Macondo: "[t]he world was so recent that many things lacked names, and in order to indicate them it was necessary to point."

Among current practitioners of conservation—a fairly recent world—it would surely be rejected as insanely irresponsible to call for a silence that would leave us only with the ability to point. But it is better to quit a language than to continue to use it to no effect or, worse, a self-undermining effect. There is a 'war' on the environment; yes, but it is no longer rational to remain in our trenches, fighting to the end. There comes a point where it makes more sense to walk into No Man's Land in order just to sit and think [43]. It at least puts us in touch with our humanity. It beats writing obituaries for the overwhelming number of casualties represented in quantitative data of the Red List.

At the dawn of meteorology, the language that spoke properly of climate was missing and had to be invented, with a language-game that could properly represent the reality formerly articulated by opinion and superficial or superstitious analogy and metaphor. At the dawn of environmentalism, before the language of the Brundtland report, there was no lack of language that represented life properly, by speaking of it according to the grammar that makes it possible to describe what living beings are, do, or have, according to their forms of life: the language of Darwin and all naturalists, but also of everyday language, that of the natural history of life forms. Again, there is irony in the flood of the new language-games, e.g. of sustainable development, among others; it should have been resisted and, moreover, it was not necessary in the first place. Perhaps now that it is so prevalent, the language could have a role to play as a secondary value that might supplement reasons grounded in life form language. But the practice of conservation of life forms must be guided primarily by the logic of a language we already know, and from which there was never any need to depart. The conservation of species threatened by extinction is not in need of words, it has them, and a grammar that guides their assembly in a systematically, logically clear way, while also implicitly containing the standard of the autonomous flourishing of any living thing.

As cited in Chap. 2, Wittgenstein suggests that a role of philosophy is to overcome the spell of language on our comprehension [35, §109]. The dominant language-games in conservation are bewitching us. The confusion is clarified, we saw, through a 'grammatical investigation' [35, §90], alert to the logic of representation. In the meantime, we must first quit the bewitching language-games and, if necessary, remain silent, or, at most, carefully 'point.' Gratefully, we can also explore the power of a language we all already know, the one that describes the natural history of life. It is only a question of recovering this language-game and facing up to its limits too when we encounter them.

Box 6.1: Overwhelming Concepts

The environmental movement has generated a full menu of language-games, and iconic terms: 'the environment' being one of them (Chap. 6). The aim or the hope has been to enlarge our perspectives and attitudes beyond human-centeredness. For example, "Earthrise" is the title of a photograph taken in 1968 by the Apollo 8 mission. It shows the appearance of the Earth on the lunar horizon. Of this image, the poet Archibald MacLeish wrote:

> To see the earth... small and blue and beautiful in that eternal silence where it floats, is to see ourselves as riders on the earth together... brothers who know now they are truly brothers [44, front page].

Earthrise became the most influential of environmental images; its extraordinary perspective on the Earth has inspired and given impetus to the global environmental movement [45, 46]. Carl Sagan said of Earthrise:

> To me, it underscores our responsibility to deal more kindly and compassionately with one another and to preserve and cherish that pale blue dot, the only home we've ever known [47, p. 21].

In contrast, the philosopher Bruno Latour contends that the idea of a terrestrial sphere disorients rather than inspires human beings, and distances them from the notion of a common home [48, 49]. He argues that the human being does not live in "the sphere" but in a tiny proportion of space that he calls the Critical Zone. Human beings tend to take care of their tiny redoubt in the Critical Zone and to represent the world from, and within, the Critical Zone, which, in the framework of the "blue planet," disappears.

Unfortunately, Latour's view is confirmed by how humanity continues to mistreat its only home. As he also once pointed out in a lecture about our idea of 'nature,' the Earth's unfathomable breadth overwhelms the capacity of our language to evaluate the threat posed to the forms of life. The concept of nature, like the concept of biodiversity, has become a catch-all that lumps together species, spaces, and processes in a representation far removed from the language of the representation of life (see Chap. 6).

References

1. Kay, L.E. 2000. *Who Wrote the Book of Life? A History of the Genetic Code*. Stanford: Stanford University Press.
2. Thompson, M. 2008. *Life and Action: Elementary Structures of Practice and Practical Thought*. Cambridge: Harvard University Press.
3. Asoulin, E. 2016. Language as an instrument of thought. *Glossa* 1 (1): 46. https://doi.org/10.5334/gjgl.34.

4. Schultz, K. 2015. Writers in the Storm: How weather went from symbol to science and back again. *The New Yorker*, November 23.
5. Trivers, R.L. 2002. *Natural Selection and Social Theory. Selected Papers of Robert Trivers*. Oxford: Oxford University Press.
6. Soulé, M.E. 1985. What is conservation biology? *BioScience* 35 (11): 727–734.
7. Kareiva, P., and M. Marvier. 2012. What is conservation science? *BioScience* 62 (11): 962–969.
8. Constanza, R., J.H. Cumberland, H. Daly, R. Goodland, and R.B. Norgaard. 1997. *An Introduction to Ecological Economics*. Boca Raton: CRC Press.
9. Beirne, P. 2014. Naming animal killing. *International Journal for Crime, Justice and Social Democracy* 3 (2): 49–66.
10. Albrecht, G. 2005. Solastalgia: a new concept in health and identity. *PAN: Philosophy Activism Nature* 3: 41–55.
11. Monbiot, G. 2017. Forget 'the environment': we need new words to convey life's wonders. *The Guardian*. https://www.monbiot.com/2017/08/11/natural-language/. Accessed 11 Mar 2023.
12. Thibodeau, P.H., and L. Boroditsky. 2011. Metaphors we think with: The role of metaphor in reasoning. *PLoS One* 6 (2): e16782.
13. Flusberg, S.J., T. Matlock, and P.H. Thibodeau. 2017. Metaphors for the war (or race) against climate change. *Environmental Communication* 11 (6): 769–783.
14. Drews, S., and M. Antal. 2016. Degrowth: A "missile word" that backfires? *Ecological Economics* 126: 182–187.
15. Grolleau, G., N. Mzoughi, D. Peterson, and M. Tendero. 2022. Changing the world with words? Euphemisms in climate change issues. *Ecological Economics* 193. https://doi.org/10.1016/j.ecolecon.2021.107307.
16. Zeldin-O'Neill, S. 2019. 'It's a crisis, not a change': the six Guardian language changes on climate matters. *The Guardian*, October 16. https://www.theguardian.com/environment/2019/oct/16/guardian-language-changes-climate-environment. Accessed 11 Mar 2023.
17. Johns, D., and D.A. DellaSala. 2017. Caring, killing, euphemism and George Orwell: How language choice undercuts our mission. *Political Science Faculty Publications and Presentations* 61. http://archives.pdx.edu/ds/psu/21103.
18. Herrera, C.M. 2008. Glosario irónico de eufemismos que empiezan por 're'. *Quercus* 271: 6–7.
19. Farrow, K., G. Grolleau, and N. Mzoughi. 2021. 'Let's call a spade a spade, not a gardening tool': How euphemisms shape moral judgement in corporate social responsibility domains. *Journal of Business Research* 131: 254–267.
20. Whitmarsh, L. 2009. What's in a name? Commonalities and differences in public understanding of "climate change" and "global warming." *Public Understanding of Science* 18 (4): 401–420.
21. Campagna, C., D. Guevara, and B.J. Le Boeuf. 2017. Sustainable development as *deus ex machina*. *Biological Conservation* 209: 54–61.
22. Du Pisani, J.A. 2006. Sustainable development—historical roots of the concept. *Environmental Sciences* 3 (2): 83–96.
23. Report of the World Commission on Environment and Development: *Our Common Future* (Brundtland Report). 1967. Transmitted to the General Assembly as an Annex to document A/42/427—Development and International Co-operation: Environment. http://www.ask-force.org/web/Sustainability/Brundtland-Our-Common-Future-1987-2008.pdf. Accessed 11 Mar 2023.
24. Mark, J. 2013. Naomi Klein: "Big green groups are more damaging than climate deniers." *The Guardian,* September 5.
25. Ball, P. 2011. A metaphor too far. *Nature*. https://doi.org/10.1038/news.2011.115.
26. Lewontin, R.C. 2001. In the beginning was the word. *Science* 291: 1263–1264.
27. Pauwels, E. 2013. Mind the metaphor. *Nature* 500: 523–524.
28. Cafaro, P. 2015. Three ways to think about the sixth mass extinction. *Biological Conservation* 192: 387–393.
29. Hendricks, R. 2017. Communicating climate change: Focus on the framing, not just the facts. *The Conversation*, March 5.

30. Geddes, L. 2022. Mushrooms communicate with each other using up to 50 'words', scientist claims. *The Guardian,* April 6.
31. Adamatzky, A. 2022. Language of fungi derived from their electrical spiking activity. *Royal Society Open Science.* https://royalsocietypublishing.org/doi/10.1098/rsos.211926. Accessed 11 May 2023.
32. Sutherland, W. 2003. Parallel extinction risk and global distribution of languages and species. *Nature* 423: 276–279.
33. Diamond, C. 2003. The difficulty of reality and the difficulty of Philosophy. *Partial Answers: Journal of Literature and the History of Ideas* 1 (2): 1–26.
34. Cafaro, P., and R. Primack. 2014. Species extinction is a great moral wrong. *Biological Conservation* 170: 1–2.
35. Wittgenstein, L. 2009. *Philosophical Investigations* (trans: Anscombe, G.E.M., Hacker, P.M.S., and Schulte, J. Revised fourth edition by Hacker P.M.S., and Schulte J.). West Sussex: Wiley-Blackwell.
36. Norgaard, R.B. 2010. Ecosystem services: From eye-opening metaphor to complexity blinder. *Ecological Economics* 69: 1219–1227.
37. Klemperer, V. 2013. *The Language of the Third Reich: LTI—Lingua Tertii Imperii: A Philologist's Notebook.* London: Bloomsbury.
38. Crist, E. 2007. Beyond the climate crisis: A critique of climate change discourse. *Telos* 141: 29–55.
39. Mccallum, M.L., and G.W. Bury. 2013. Google search patterns suggest declining interest in the environment. *Biodiversity and Conservation* 22 (6–7): 1355–1367.
40. Raymond, C.M., G.G. Singh, K. Benessaiah, J.R. Bernhardt, J. Levine, H. Nelson, N.J. Turner, B. Norton, J. Tam, and K.M. Chan. 2013. Ecosystem services and beyond: Using multiple metaphors to understand human–environment. *BioScience* 63: 536–546.
41. Neilson, A. 2018. Considering the importance of metaphors for marine conservation. *Marine Policy* 97: 239–243.
42. Le Guin, U.K. 1985. She unnames them. *The New Yorker,* January 13.
43. Campagna, C. 2013. *Bailando en Tierra de Nadie: Hacia un Nuevo Discurso del Ambientalismo.* Buenos Aires: Del NuevoExtremo.
44. MacLeish, A. 1968. A reflection: Riders on the earth together, brothers in eternal cold. *The New York Times,* December 25. https://www.nytimes.com/1968/12/25/archives/a-reflection-riders-on-earth-together-brothers-in-eternal-cold.html
45. Davenport, C. 2018. Earthrise: The stunning photo that changed how we see our planet. *The Washington Post,* December 24. https://www.washingtonpost.com/history/2018/12/24/earthrise-stunning-photo-that-changed-how-we-see-our-planet/. Accessed 11 Mar 2023.
46. Myer Boulton, M., and J. Heithaus. 2018. We are all riders on the same planet. *The New York Times,* December 24. https://www.nytimes.com/2018/12/24/opinion/earth-space-christmas-eve-apollo-8.html. Accessed 11 Mar 2023.
47. Sagan, C. 1994. *Pale Blue Dot: A Vision of the Human Future in Space.* New York: Random House. https://www.pdfdrive.com/pale-blue-dot-a-vision-of-the-human-future-in-space-e176236257.html. Accessed 11 Mar 2023.
48. Latour, B. 2016. Why Gaia is not the Globe. https://cas.au.dk/futures/bruno-latour. Accessed 11 Mar 2023.
49. Latour, B. 2016. On not Joining the Dots: Land... Earth... Globe... Gaia. Radcliffe Institute. https://www.youtube.com/watch?v=wTvbK10ABPI. Accessed 11 Mar 2023.

7 Introduction to the Language of Extinction

To admit that species generally become rare before they become extinct—to feel no surprise at the comparative rarity of one species with another, and yet to call in some extraordinary agent and to marvel greatly when a species ceases to exist, appears to me much the same as to admit that sickness in the individual is the prelude to death—to feel no surprise at sickness—but when the sick man dies to wonder, and to believe that he died through violence.
C. Darwin [1, p. 306]

We have seen that threats to nature tend to be identified and evaluated according to the standards implicit in the language that describes them. If the standard of goodness is beauty, an oil spill, such as that caused by the Deepwater Horizon in 2010, is an unthinkable catastrophe, as reflected, perhaps surprisingly, in the August 2010 issue of the fashion magazine *Vogue Italy* [2]. We tend to identify and address only the causes of the provoked extinction that the standards of dominant language-games allow us to understand. The economic-instrumentalist language-game illustrates this when it represents extinctions that we provoke as equivalent to those that occupy paleontologists: namely, those due to impersonal, natural forces. The logic of economic-instrumentalist language, in practice, underwrites concern for other species only in terms of how they are instrumentally valuable to us. The Brundtland Report says: "... all species should be safeguarded to the extent that it is *technically, economically, and politically feasible*" [3] (Chap. 1, italics ours). Predictably, efforts will not be invested in species judged useless, or harmful, or uninteresting to those in charge of giving thumbs up or down. The Brundtland Report uses the term "inherent value," in a sense similar to "intrinsic value," but even so, its language-game is so predisposed to the priority of the instrumental value of nature for us that the variation in terms makes little difference. The Report is the source and paradigm for the 'nature-for-development' approach, its language has evolved [4] into a political discourse used by every government that adheres to the

C. Campagna, D. Guevara, *Speaking of Forms of Life*, Fascinating Life Sciences,
https://doi.org/10.1007/978-3-031-34534-0_7

improbable notion of infinite economic growth. In the face of the threat of provoked extinctions, the Report is predisposed toward instrumentalist solutions, and to accepting the loss of species as 'inevitable.' Therefore, the compromised strategy seems to be to advocate for a shift from indiscriminate provoked extinctions to selective ones.

In sum, the dominant practice of conservation justifies provoked extinctions on the basis of the requirements of development and as a by-product of behaviors that are not even related to human needs. Little to no attention is given to the logic and values of the language used to represent extinctions.

This chapter discusses the notion of extinction through reflections on the relation between extinction and death, and the difference between the 'population-based' conception of extinction and the life form perspective.[1] It is proposed that a conservation practice that attempts to stop human-induced extinction has to understand that it faces a 'difficulty of reality' in Cora Diamond's sense, introduced in our last chapter,

> . . . the sense of a difficulty that pushes us beyond what we can think. To attempt to think it is to feel one's thinking come unhinged. Our concepts, our ordinary life with our concepts, pass by this difficulty as if it were not there; the difficulty, if we try to see it, shoulder us out of life, is deadly chilling [5].

The difficulty is presented not by extinction as a general phenomenon in nature but by the fact that it is brought about by acts involving human will. The phenomenon in nature is a well-understood reality, where extinction plays a crucial role in the generation of the diversity of life forms by natural selection. Extinctions unnecessarily provoked by human will are another matter entirely [6], one that could bring about the end of life forms as we know them and therefore something which we have hardly begun to comprehend.

The Word "Extinction"

"Extinction," in an older meaning, applies to a fire that is smothered or extinguished. The flame of a candle, a forest fire, or a volcano is *extinguished.* The term now also indicates the end of a process, such as the end of a life. In Spanish, for example, "*extinto*" refers to a dead person, as in the phrase: "*honrar la memoria del*

[1]Conceptual differences between the notions of species-populations and that of life forms will be developed in Chap. 17.

extinto. . . ." Metaphorically, death extinguishes life like a flame. Applied to species, the term extinction began to be used at the end of the eighteenth century, with the paleontological disciplines.[2] In the language of paleontologists, it represents the species whose members are no longer found among the living.[3]

The language of species extinction and that of individual death differ. But Darwin relates death and extinction (see header quote) in an analogy meant to illuminate the gradualist perspective of evolution: extinction is reached through successive stages of loss of abundance, rather than through catastrophes. Just as a death preceded by a long agony is not necessarily due to violent causes, neither is the extinction of a species necessarily due to catastrophic causes; it may be the result of successive stages of a population becoming more and more rare.

The perspective that dominates the way practicing conservationists represent extinction today is found in talk of 'populations,[4]' which are essentially just sets of individual living things organized according to parameters of interest to ecologists, geneticists, etc. This talk is characterized by terms like 'abundance' and 'rarity,' and also 'growth,' 'size,' 'trend,' 'mortality,' 'natality,' and 'fecundity'—as applied to sets of individuals. Extinction is an extreme on a continuum of varying abundance or rarity. The details of individual life—what living things do, for example—are not necessary to this way of representing extinction. For this reason, as we will argue, population language cannot even approximately represent provoked extinction. It is remarkable that it has been so generally accepted as appropriate; it is like being satisfied with representing genocide by human demographic graphs and parameters. Demographic data will support claims of threat or extinction, as of genocide, both involving large numbers across time and place, but the point we are emphasizing is about the quality of the confronted reality, not represented by quantitative analysis. Also, the point is not that there is moral equivalence between genocide and provoked extinction, but about a similarity in the difficulty of coming to grips with the reality we confront.

Provoked extinction involves the death of every single representative of a life form. If the language leaves out the representation of each of these individual deaths,

---

[2]Count Buffon, one of the founders of paleontology, used it in 1775. In 1860, paleontologist Richard Owen uses it in the title of his book: *Palaeontology* [sic.] *or a systematic summary of extinct animals and their geological relations*. Edinburgh: Adams and Charles Black.

[3]The term "extinction" is also used to represent the disappearance of a human language or dialect (see Chap. 6, reference [31]). The extinction of a language means that it no longer has any users: no one speaks it as a native speaker. Languages are referred to as 'threatened with extinction,' with the threat defined by a drop in the number of people using the language (a population perspective). A 'dead language' is one that no one speaks as a native speaker, although it may still be in use as a secondary language: like Latin is in some academic, professional, or religious contexts.

[4]Ecologists understand species as composed by "populations," often distributed in a wide geographic range. They rightly point out that the extinction of populations is far more common than that of a species, a fact that conservationists should consider [7]. In the present context, we do not ourselves distinguish between the terms "species" and "populations," as our focus here is on the language-game that represents living things following a quantitative approach, which shares the same grammar when discussing species or populations.

it will not be able to describe the 'death of the whole.' Also, it must represent the fact that the individual deaths are the consequence of human choice and action, as with the passenger pigeon. It is here that language must overcome a difficulty of reality, in Cora Diamond's sense: the difficulty of capturing the full significance of the 'death of the whole kind,' through the killing of each and every individual bearer. One seems to transcend the other—the whole is not just individuals making up the collection or population—and yet what has been done to the whole must be understood through what has been done to each individual.

So far, in accepted conservation discourse, the difficulty has been 'solved' by deflection, i.e., by language-games that fail to confront the difficulty of this reality, at least as regards other species. When it comes to human beings, we do have some language that does justice to both the representation of the individual and the whole. A genocide can be described with quantitative data appropriate for population analysis, but it is also commonly described through the suffering of individuals and their identity as a 'people,' a whole race or culture or ethnicity. These latter concepts are not always clear, but the intended logic is: the intention is to represent a certain kind of logical or grammatical relation between individuals and their kind, in a sense not reducible to talk of population or collections of individuals, or even mass killing. Profoundly gifted writers, like Svetlana Alexievich and Primo Levi [8, 9], have perhaps given us language to confront the difficulty of the reality of a concentration camp or a blown nuclear plant. In our opinion, the tragedy of a provoked extinction still awaits equivalent articulation.

The extinction of a species-population is the extinction, in the sense of death, of all its individual representatives. Therefore, the life history of each individual should matter. Moreover, the representation of the individual should be the entry point for representing the extinct species. However, population language refers to the life of the individual only indirectly and generally, as members of a set characterized by quantitative parameters, e.g., parameters that are constructed by integrating individual events, such as mortality and birth rates. Population language does not need to express what living is about in order to represent extinction. There is no need to refer to the grammar of life forms, or its implicit standards, to understand the statement "the passenger pigeon is extinct," when understood as saying something about a species-population. In contrast with life form language, there is no implication about whether a dead passenger pigeon is a good representative of the form or not. There is no explicit concern with the satisfaction, or not, of the necessities of the form. In the same way, no living thing, as an individual, satisfies 'abundance' or avoids 'rarity.' The qualities of the whole, in case of a population as a collection of individuals, do not take into account what is involved in the development of the individual life.

To recapitulate, the term "extinction" went from expressing the extinguishing of an individual flame or the like, to an individual's death, and, now, to representing a population of zero size, the end of the existence of a set or collection. In this last representation, there is no need to represent the life of the individual, nor to consider causes—including avoidable, intentional choices and actions—as relevant to the extinction. Darwin at least thinks still in terms of an analogy between individual death and extinction. For current conservation practices, individual death may have no relevance at all, not so long as population parameters remain within what is considered statistically acceptable. That is the framework of the IUCN Red List. Today, such thinking dominates talk of a provoked extinction too. Moreover, the effects of underlying human will do not enter into the computation of the population parameters that guide value judgments. 'Extinction' is now a concept that, while rooted in the life process of particulars or individuals, has been stripped of the grammar of dying. The 'flame that goes out' is now represented by an empty collection or set.

The Last Individual

The IUCN Red List categories illustrate how population language is typically applied to extinction [10]. The essential data underlying each category relate to population "size" and "trend." A threat (overexploitation, pollution) affects the trend in a negative way: e.g., populations become "rarefied." Interventions seek to correct the 'wrong,' i.e., the fall in the size of the aggregate. The individual, obviously, underlies these parameters, but the language does represent it directly. The exception is the category 'Extinct.' A species is extinct when "...there is no reasonable doubt that the last extant individual has died out." The definition acknowledges the reality of the particular in this case: the case of the last of its kind and its death.

The language-game of 'the death of the last individual of its kind' is not the same as that of the whole, or the other members of the whole. For example, it seems fitting to name and memoralize the last of its kind, as we have done with Martha, the last passenger pigeon, who died in the Cincinnati Zoo around September 1, 1914. The last giant tortoise of the Galapagos, named Lonesome George, lived more than a century and died on June 24, 2012. For some species, there is a record of the cause of death of the last representative in natural conditions: "The last known thylacine to be killed in the wild was shot in 1930 by Wilf Batty, a farmer from Mawbanna in the state's northwest" [11]. But this language is not relevant to the understanding of population parameters, nor does it describe any individual that might have been 'spared' in captivity. Except for humans, we do not routinely keep track of the deaths of individuals of any wild species; we do not take note of them the way we do the last of their kind. The practical impossibility of doing so is obvious. In general, then, only two grammars of conservation biology apply to expressing provoked extinction: the grammar of populations and the grammar of the last individual of its kind. Neither represents what living things cease to be, have, or do when their kind is eliminated by human causes.

The Human Double Standard

Demographic-population language is applied to human beings in contexts such as public health, life insurance, population censuses. But this language is not in general at the expense of an individual person's death. For human beings, the difference between individual and population matters. The personal language that is used only for the last of its kind, when dealing with other species, is as common to human life as the obituaries of the daily paper, and not the historic rarity of Martha or Lonesome George.

The language that deals with human death also differentiates between individual and mass death. There are even differences among types of mass death: a genocide is described differently from the casualties of war or a pandemic, although, demographically, they are subsumed under the equivalent parameters. Deaths by heart attack, accident, war, abuse in concentration camps, or contagious disease have their particular language-games, as do those of individual deaths. Suicide is represented differently from that of one who dies fighting off a terrible illness, or by accident or murder.

We see, then, that so far as *our* life form goes, language is sensitive to a wide variety of distinctions in death: from individuals to masses and kinds, and to the context and causes of death, whether of individual or whole. Humans value, and take pains to represent, the death of individuals of their form of life, and that of a few other life forms, which they have made the target of law-based practices and animal welfare. These nuanced representations are not applied to the great majority of natural species, but instead population language dominates, with resulting anonymity of the individual. Our point is not to name, or write obituaries for them all, but to describe extinction with a language that expresses what life was about for the lost individuals, in terms of the great variety of extinguished forms they once bore. Rate of mortality, or fecundity, are poor descriptors of the birds of paradise.

The Implications of Evaluative Judgment Based on the Language of Population

Population grammar represents states of affairs as good or bad in terms of a population threshold in which the individual participates as a number. 'Good,' for example, is a stable or growing group size, while the passage from abundance in number to rarity is usually evaluated as 'threatening.' Consequently, as long as population parameters remain in the 'good' range, it seems permissible to cause the death of any number of individuals. This is how industrial fisheries have been managed. In fisheries biology, for example, the particular individual is irrelevant; it is the group data that dictates whether you can extract individuals from the population (understood here in terms of their *biomass*!). What is good depends on 'extracting' individuals in an economically productive way. The grammar of this 'good' allows also for 'removal' of individuals of particularly attractive species through, e.g., trophy hunting, even when the population is not abundant. The justifications may be demographic or economic: e.g., a hunter pays to be able to kill an individual that no longer 'contributes' to the growth of its population, or even when the individual may still be able to contribute to growth, killing it is still alright so long as the population is considered 'healthy.' The individual is thus canceled out by the language of the whole or the 'benefits' associated with its death.

The aim of a practice of the conservation of life forms is to reverse this disappearance of the individual from the story, integrating individual natural history into the representation of the state of affairs of the whole kind. The fundamental reason for practicing conservation on this view lies in what best protects what an individual living thing is, has, or does, according to its form of life. It is a practice that does not abandon the individual for the sake of quantifiable parameters of a whole in which individual existence is marginalized. Nor does it try to recognize, or act on behalf of, the individual with the language of 'rights' or 'welfare' borrowed from the human form of life, as tends to happen now, especially in last-ditch efforts to save the last of their kind. A practice guided by the language of the conservation of the forms of life respects the autonomy of the forms, and represents and evaluates provoked extinction according to the standard set by the necessities of the relevant forms, and does not appropriate standards from other life forms. For the passenger pigeon, it is important that it reproduced in large colonies. The provoked extinction erased precisely the possibility of satisfying this fundamental need, among countless others, without justification based on the natural needs of our own form (although there may be rationalizations based on supposed economic necessities, etc., but none of which could justify killing to extinction). For this reason, it is unacceptable by the standard of goodness set by the form of life.

The term 'provoked extinction' is already an evaluation of a state of things: 'provoked' associates it with human choice and action and, thus, responsibility. It is, as we have said, a forcing of the loss of a species through human will. Many current conservation practices attempt to redress the threat of provoked extinction as though what is wrong with the threat is that it might lead to zero population size. That, in turn, might be thought bad for any number of reasons. But no current conservation practice sees what is wrong with provoked extinction as lying exclusively, or even primarily, in our making it impossible for the natural goodness of a life form to manifest itself any longer. No one today joins the discourse of extinction insisting on representing and evaluating it with the language whose grammar reveals the needs that each extinguished individual once satisfied as a flourishing member of its form. But, we insist, this is the language that best represents, and allows us to evaluate, a provoked extinction, if what we care most about is what is good for living things, according to all their diversity of forms. Interventions based on this language, obviously restricted to stages prior to extinction, are first and foremost based on concern for the threatened life form and the natural goodness manifested in its bearers.

Extinction of a Form of Life

'Form of life' is a fundamental term of this work (buttoned down in Chaps. 10 and 17). The corresponding concept is articulated by all the natural historical descriptions that differentiate living beings from each other (and from the non-living) according to what they have and do to satisfy the requirements of their form. The individual and the form are integrated: an individual creature lives by exemplifying the recipe for living for its kind. The form of life of *a* passenger pigeon is given by representation of '*the* passenger pigeon,' the latter understood as referring to the kind. The representation of the specifics of a form or kind is arrived at through experience, by applying the general schema of the form-bearer relationship to the specific shape of the lives of individuals bearing it, and each experience improves and expands our grasp of the specific shape of a life. We come to learn that a particular individual thing here and now—a caterpillar, say—is experienced for what it is only as we understand our experience to be presenting us with stages of a broader developmental context leading to a butterfly. And so it is, with all living things as opposed to rocks or metals or any other inanimate objects. The general relationship of form to particular bearer is mutually interdependent and necessary.

There are superficial similarities between the concept of species as used in population biology and that of the language of life forms.[5] Much of what would describe the extinction of a species as population is implied by the extinction of a life form. A chief concern in this section has been the following difference: according to biological species as populations, extinction is an extreme in population rarity and, according to the concept of life forms, extinction is that but also the permanent loss of the possibility of innumerable instances of a certain kind of creature's satisfying the necessities of its life. The examination of the language of life forms uncovers a particular logical structure that allows, simultaneously, representation of the individual and the form it bears (Chap. 10). In describing a living thing, we illuminate the shape of its life. In ordinary language, judgments about the health of a creature—e.g., about whether it is vigorous, sick, or wounded—are only possible by comparing it to its form. Unless it is on purpose, robotic movement in a human being is defective because it is not described by the natural history of human behavior; it is not natural human movement. There is nothing about robotic movement, with its right angles and abrupt interruptions, that describes human behavior like "Human beings walk upright, in fluid, continuous motion." According to this language-game, 'defect,' in its most general sense, refers to something in the creature that falls short of the requirements of its form; it is not to be confused with the discriminatory or pejorative use of the term (see Box 7.1). We are all defective in this sense, since no creature fulfills all the necessities of its form. Nobody is perfect.

A form of life is extinct when there are no longer any autonomous living bearers of it. Its extinction implies an indeterminately long list of natural needs—and (thus) potential natural goodness—that can no longer be satisfied or instantiated. Extinction must be understood according to what the language of the death of all bearers of a form involves, and thus some understanding of the loss in terms of all the natural goodness no longer possible. If each individual counts in the representation of the

[5] In this essay, we sometimes use "species" as a synonym for "form of life" ("life form"), but there are fundamental differences between the ways the term "species" is most commonly used in biology and that of "form of life" as we use it. Both express concepts that are essential and each follows its own grammar. In Chaps. 10 and 17, the two concepts are shown to differ in a fundamental respect. In Chap. 10, we present Thompson's arguments that life is an indispensable logical category, similar to other fundamental logical categories, such as quantity, quality, or substance.

form, then each individual should matter to the representation of the extinction of its form. As we will see (Chap. 10), the notions of individual and form are inextricably linked. But the language-games played with "population" do not necessarily express this link, since they are concerned to circumscribe the parameters of a set, according to a wide variety of criteria hardly ever focused on the life form needs of the individuals in the set. For example, the population parameters do not typically describe how reproduction or foraging occurs; they submerge these and other aspects of the form of life under birth or mortality rates, among other quantitative measures. This language might have some capacity to help with the practice of preserving life forms, but it is to provoked extinction as geometric language is to ordinary human movement.

In sum, to speak of species 'having gone extinct,' or to focus on the qualities of the *last* individual, or to assess as good or bad primarily in terms of population decline, departs from life form language; it is talk that alienates us from what it means to live, what it is to *be* a living thing, for in this type of population language what matters is whether the individual is simply alive and thereby adds to the 'effective' size of its population. Parameters, such as the fertility rate, might relate to the satisfaction of natural necessities, but the primary concern with populations is not with those necessities; rather the concern is, as noted early in this chapter, with statistical generalities like trend, abundance, or rarity. Nor do the developmental stages of a life cycle matter. Age matters, but only for life tables differentiating, for example, adults from juveniles, in the case of vertebrates. The particular necessities that each age category may satisfy are not part of the table and do not add necessary information for evaluative judgment. To die is to cease contributing to the quantitative parameters that define what is 'good' for the population.

In the conservation of the forms of life, the 'death of the whole' is inextricably connected to the death of each individual. In current English, unlike Spanish or some other languages, 'extinct' only applies to kinds, not particular individuals (although we can speak loosely of an individual's being dead as being 'extinct'). But either way, the grammar of natural life requires a representation of the organic, mutually interdependent relationship between individual and kind, and therefore requires that we understand the irreversible loss of the one as inextricably tied with the irreversible loss of the other. There is a sense in which, in the conservation of life forms, the loss of the last of its kind is no different than the loss any other bearer of the kind. Our threat to any living thing is as much a threat to the form as the threat to the last bearer is. To have no concern for the individual is to have no concern for its form of life. This follows from the logic or grammar of life; it is not a matter of choice. Extinction makes the affront irreversible and permanent for the whole, just as the killing of one individual does for the particular, and so in a sense is worse.

In the language of life forms, understanding what it means to extinguish a life form involves understanding what it means to conserve it, what it means to keep intact and functioning the values implicit in the natural history of the species, as exemplified in the individual representatives. Conservation of life forms is conservation of living representatives and of the conditions for them to satisfy the needs of their form. The language of life forms implies that any conservation practice that prevents the extinction of the form is necessarily also a practice of conserving each individual representative out of concern for its ability to satisfy the indispensable needs of its autonomous life.

The practice of the conservation of life forms is necessarily the practice of the conservation of living representatives of the form because, as will be repeated many times in this work, form and individual bearer are necessarily interconnected. A helpful comparison might be with the relationship between respect for humanity and respect for individual human beings; there is a sense in which there cannot be one without the other. In this way, the practice we propose differs diametrically from a practice derived from the principles governing, for example, the Red List categories. Those principles ignore the fact that when countless creatures are 'taken,' as in industrial fishing, or even when a single one is taken, as in trophy hunting, it is an affront to the form. The disrespect in the unnecessary killing of single individual creature is disrespect for the form of life in that individual, but it affects the quantitative measures of a population only in a negligible way.

Reconversion of Meaning

Summarizing: Under the language of life forms, the meanings of "extinction" are integrated; the life form becomes extinct as the representatives become 'extinct' (*extinto*) or extinguished (in the sense that they die). Every particular counts, not just the last one; in fact, the life of the last individual of a threatened life form is very *defective*, considering the impossibilities of satisfying needs when there are no others of its kind.

As the language of life forms does not subsume the individual under parameters of the whole, the notion of extinction can be explicated at the level of each particular '*extincto*,' each dead particular. The practical impossibility of expressing the innumerable qualities of each individual eliminated by a lethal act of human will is obvious, just as it is impossible to write obituaries for each and every dead representative of a life form. But what is important is to understand what would be the appropriate language for expressing provoked extinction of a life form, with a view to the life of the individuals extinguished. And this is given by the extinguishing of natural goodness, and its very possibility, in the permanent loss of individual bearers of a form.

In general, the prevailing language of conservation practice today does not articulate threats in terms of natural goodness or badness. Interventions to address the threats of provoked extinction are based on judgments not necessarily connected with the standards of the shape of a life. No living thing reproduces to satisfy a standard of 'abundance.' Prevailing judgments thus neglect the immensity of what may be wrong, harmed, or lost in countless repetitions of extinguishing individual life; they neglect that which we have proposed essential to any representation of the extinction of form or species. It is the loss or harm in this unnecessary, relentless natural badness that comes closest to representing the provoked extinction. It took hundreds of millions of passenger pigeons—each satisfying innumerable vital needs—extinguished by voluntary human acts, to arrive at the extinction of the form. Current conservation discourse does not discuss extinction in this way, but, as we have urged, language does nevertheless exist for representing this essential element of provoked extinction, even if life form language does not allow us to fully comprehend the difficulty of this reality that we willfully bring about. It may be a reality that continues to escape language, as genocide once was.

Implications for Human Natural Goodness

How are we to properly evaluate the human act that extinguishes life forms, or the human agents behind the act? We will devote entire chapters to this question (Chaps. 21 and 22). We have seen that with population language—which generally ignores the intimate link between threat to a particular creature and the form it bears—it might even be reasonable to *bring about* a provoked extinction, most obviously in cases of 'harmful' species. Indeed, as we saw, the managerial idea of exerting discretion over which species should be saved from extinction is suggested by the Brundtland Report. The terms 'loss,' 'harm,' and 'impairment' are played according to various alternative language-games when the standards of evaluation are not those of the life form. For example, an extinction might be a 'loss' of an opportunity for profit, or a 'harm' or 'impairment' to the development of an industry (see [12]). The standards of life forms imply evaluative judgments entirely opposed to these, logically incompatible with them.

On the theory being developed in this essay, human natural goodness involves human recognition of, and regard for, the autonomous needs of other forms of life. As we will argue in more detail later, human interference with those needs is justified only when necessary to the satisfaction of our own life form needs (e.g., the need for self-defense, food, or shelter).[6] Then, on the view developed here, a provoked extinction transgresses values implicit in certain natural historical judgments: e.g., the value implicit in the judgment that human beings recognize and understand the natural good of other forms of life (inasmuch as they understand the relationship between life form and creature bearing it). Then, in understanding natural goodness across forms, humans understand that the reasons for respecting one instance of it are in general the same as the reasons for respecting other instances. Differential treatment of the satisfactions of the needs of one form (ours say) over another requires a justification in terms that are consistent with general respect for natural goodness.

[6]Human population expansion has the result that the satisfaction of even the most fundamental of the necessities of the human form may represent a threat of extinction. As we note in the last chapter, on objections, this is one of many difficult ethical issues we must eventually confront in the specific application of our theory (see Chap. 26). Still, it is worth noting that many of the examples of human action that cause extinction are not directly related to population issues. Fishing discard, for one of many examples, is due to commercial practices not directly caused by the size of human population but by the quality of the demand that can be under human control.

The detailed, specific implications of this, and many other issues, for human life must be carefully worked out in another work, but in the final sections of this essay we will begin to draw out some of implications, and to answer initial objections. Moreover, there are certain things that seem clear generally. For example, although self-defense is part of the natural goodness of any life form, it does not seem that justification of the harms we might do in self-defense, etc., could be sustained when they lead to extinction. Even if the passenger pigeon caused the harm that supposedly justified killing it in 'self-defense' (consumption of seeds of a crop, for example), it is unlikely that it continued to cause such harm as the species became less abundant or rare. Yet, the killing was never effectively forbidden. Consequently, the acts that led to the passenger pigeon's extinction must have transgressed, at some point, human standards of natural goodness, because the killings to annihilation could no longer be justified by a necessity of our form. The point is generalizable to any provoked extinction, as developed further in Chap. 21.

Conclusions

A provoked extinction represented by the language of life forms is represented by (1) the innumerable killings, of bearers of the form, (2) the innumerable needs, described by the natural history of a form, which no representative any longer satisfies or ever will, (3) the innumerable standards that no longer have any application, and (4) the impossibility of a flourishing life for bearers of a particular form of life. Extinction is *not* simply a matter of reducing a population to size zero, nor a quantitative expression of certain parameters characterizing the members of a set, nor a loss of opportunity for use, nor a phenomenon common to all extinctions that have occurred during the history of life. To represent a provoked extinction with the language of life forms is to express the loss, forever, of the value of any living thing of a certain kind satisfying the needs of its kind.[7] The language of life forms thus represents provoked extinction as the result of a relentless human natural badness perpetrated on the kind, through its extinguishing of the creatures that once bore the now extinct life form.

The language of life forms, as we will see in Chap. 10, logically precedes, and is therefore necessary to, the possibility of any alternative language for representing extinction—such as the language of populations. It is logically prior and necessary, we will see, because one cannot represent a living thing without relating it to the form of life it bears. Understanding life requires understanding particular individual things, here and now, in the broader context of their form. Then, it is ironic that practices whose reason for being is to redress provoked extinction attempt to understand the death of all the particulars through parameters of a whole quite

[7] These arguments are similar to those based on the concept of intrinsic value, but it is a similarity that requires its own grammatical investigation in order to also keep track of the important differences. The grammatical analysis is carried out in Chap. 13.

distant from anything that could be a form of life. Such a practice cannot even begin to cope with the difficulties of the reality of provoked extinctions and the overwhelming weight that language must bear when properly representing them. It is overwhelming to consider giving expression to the inexhaustible list of losses in terms of the natural historical propositions that describe the form, and the enormity of corresponding good that was once instantiated by flourishing representatives of the form.

To declare a species extinct simply in terms of the population size being zero is a failure of representation, at the level of a Wuhan-style blunder. The shortcomings of trying to assess extinction as a result of 'unsustainable' activity are clear, as well. The alternative language is useful, perhaps, in the framework of certain biological or economic disciplines, but not for the practice of life form conservation. On the other hand, it might be possible to adopt and adapt special language already in use for some cases of mass human killing, which at least is an attempt to come to terms with the unprecedented enormity of the situation. But, again, as we know from the history of the concept of genocide, hard work is required, and it is not just a matter of coining or extending a term. New terms, or new applications, must be integrated into the 'economy' of our existing ordinary language, or else it is simply 'word inflation' so to speak: too many new words chasing around too few familiar language-games. In the case of provoked extinction, there must first of all be general acceptance, and explicit integration, of the language of natural goodness into ordinary language, initiated perhaps by the culture of conservation and a proper representation of the role of human choice, action, and responsibility, without interference from the language that erases the distinction between natural and provoked extinction. The difficulties might seem insurmountable, were it not, as we will later discuss, that the language which must be generally prioritized is already universally in use, but has been over-shadowed by language-games strategically forced to justify human secondary needs over primary.

Box 7.1: Defect

Announcing one of her main ambitions, Foot says:

> I believe that evaluations of human will and action share a conceptual structure with evaluations of characteristics and operations of other living things, and can only be understood in these terms. I want to show moral evil as 'a kind of natural defect.' [13, p. 5]

Moral evil is a natural defect that can be attributed only to creatures of the human form of life (or, if there are any, to those with rational agency like ours). Foot's moral theory, which will be summarized in Chap. 12, treats moral goodness as an instance of human natural goodness, and moral badness, therefore, as a human natural badness or 'defect:' a defect of character and will. Moral goodness and defect are close to the traditional concepts of virtue

(continued)

Box 7.1 (continued)

and vice, and we will make use of this point in Chap. 22. But the general use of "defect" in Foot's sense is quite general indeed: we find defects in this sense in creatures of all forms of life, wherever they fall short of the requirements of their form. The term "defect" in this sense is useful as a way of speaking about natural badness in general (across life forms), although it should not be confused with uses of "defect" that stigmatize people who do not fall within certain narrow-minded, cultural norms, for example. The only relevant norm is the primary standard for life forms: the satisfaction of the requirements of one's form. On this standard, nobody is perfect and human 'defect' ranges from not having the full set of 32 teeth (whether through a street fight or congenital defect) to moral viciousness.

Box 7.2: Death and Killing

> As in death, too, the world does not change, but ceases. Death is not an event of life. Death is not lived through.
>
> L. Wittgenstein [14, §§6.431, 6.4311]

Death

Wittgenstein's observation in the header quote complicates the representation and evaluation of death in the language of life forms. Presumably, on that language, "death" would signify a defect like "sickness," "wound," "coma." It would be, in this respect, the ultimate defect. The complication is that the standard used to determine natural goodness or defect applies to what a creature is, has, or does, i.e., something lived through, like illness or a wound, etc. With death we would be applying the standard to the cessation of all that. Wittgenstein's point alerts us to a difference in grammar that we will need to keep in mind.

A further complication is that it seems death can be part of the natural good. 'Adaptive altruisms' are an example: "the worker bee often dies defending the hive" is a natural historical proposition describing a necessity of the life form, for certain members of it. It is the same for males of some species of mantises or spiders, which are commonly cannibalized by the females during copulation. When worker bees die defending the hive, or male mantises are beheaded during copulation, things are as they should be according to the life form. So they seem to be part of natural goodness. Humans often react negatively to these natural phenomena. The reaction is comparable to those that have drawn objections to the theory of natural goodness (see Chap. 26): e.g., to the brutal

(continued)

Box 7.2 (continued)
fighting that occurs as part of the natural history of the elephant seal. How can such things, and especially death, be good in any sense?

What is important to recognize here is that all creatures must expose themselves to harm and death in order to satisfy certain necessities of their forms of life. This happens in human life, most nobly and dramatically, when one must choose between death and collaborating with evil, for example. An example from other life forms that could be added to the above is 'adaptive suicide' [15]. This occurs, for example, in infected colonial microorganisms, which die in order to impact the life cycle of the pathogen that parasitizes them. In some unicellular colonial microorganisms, 'biomass sacrifices' occur: under certain resource-limiting conditions, one part of the colony, the older one, dies, generating resources for the newer one [16].

The lesson is that in the language of life forms, if asked whether something is good or bad, we are able to answer only if we know something about how it relates to the relevant form. And, of course, for the vast majority of living things, the relevant form is not human life, nor therefore what would be considered intolerable in relation to that form.

We began by suggesting that death would be treated as a defect by the language of life forms. The truth is that, whether lived through or not, it must be related to its role in the natural history of the creature, before we can say whether it is naturally bad or good. As the proposition about worker bees shows, death is part of the natural history of a form. Like any other death, it is the result of all kinds of defects, breakdowns in its vital functions. But if the bee died by doing what its form requires, its death is an inevitable part of the satisfaction of its form and, thus, natural goodness.

Killing

Similar points can be made about killing. Killing can also be part of the natural good in countless forms of life. Predation is the obvious example. But there are many others. Matriphagy is a form of parental care of some invertebrates by which the mother is consumed as food by the offspring. In physiogastric reproduction, as in some lice, the females lodge their offspring in the digestive tract. There the males fertilize their sisters and die. The females continue to develop and emerge mature. In the process they kill and cannibalize the mother. Cases of 'infanticide' or 'fratricide' could also be cited as examples of the satisfaction of the needs of certain life forms.

Of course, humans also kill out of necessity, to feed themselves, for example, and thus out of natural goodness. That we harm and inflict natural badness on the animal or plant we eat is clear, but what we do is good, in keeping with the primary standard for our will and character, so long as it is done out of the need to satisfy the requirements of our form. This is in stark contrast with provoked extinction, which is often brought about through

(continued)

Box 7.2 (continued)
innumerable unnecessary acts of killing.[8] The language of life forms differentiates killing and dying as natural necessity, from the rest of the causes and justifications of death.

The crucial thing always is the standard: if we are considering whether something a creature is, has, or does is naturally good or bad, we look to the role it plays in meeting the needs of the creature's life form. Then, whatever is necessarily or unavoidably involved in that, even if it harms or kills (i.e., causes some natural badness), is naturally good. **Only for the human being is there a limit when the act of killing, due to the context in which it occurs, leads to extinction.**

References

1. Darwin, C. 1997. *The Voyage of the Beagle*. The Project Gutenberg eBook 944. https://www.gutenberg.org/ebooks/944. Accessed 11 Mar 2023.
2. Combes, T. 2010. Vogue Italia runs a controversial oil spill-inspired editorial. *Refinery 29*, August 5. https://www.refinery29.com/en-us/oil-spill-fashion-photoshoot-from-vogue-italia. Accessed 11 Mar 2023.
3. Report of the World Commission on Environment and Development: *Our Common Future* (Brundtland Report). 1967. Transmitted to the General Assembly as an Annex to document A/42/427—Development and International Co-operation: Environment. See: http://www.ask-force.org/web/Sustainability/Brundtland-Our-Common-Future-1987-2008.pdf. Accessed 11 Mar 2023.
4. Hajian, M., and S.J. Kashani. 2021. Evolution of the concept of sustainability. From Brundtland Report to sustainable development goals. *Sustainable Resource Management*. https://doi.org/10.1016/B978-0-12-824342-8.00018-3.
5. Diamond, C. 2003. The difficulty of reality and the difficulty of philosophy. *Partial Answers: Journal of Literature and the History of Ideas* 1 (2): 1–26.
6. Monbiot, G. 2017. Forget 'the environment': we need new words to convey life's wonders. *The Guardian*. https://www.monbiot.com/2017/08/11/natural-language/. Accessed 11 Mar 2023.
7. Ehrlich, P.R., and G.C. Daily. 1993. Population extinction and saving biodiversity. *Ambio* 22 (2/3): 64–68.
8. Alexievich, S. 2006. *Voices from Chernobyl: The Oral History of a Nuclear Disaster*, Trans. K. Gessen. New York: Picador.
9. Levi, P. 1979. *If this is a Man*. Penguin, London.
10. IUCN. 2012. Red list categories and criteria. Version 3.1, Second edition. https://portals.iucn.org/library/node/10315. Accessed 11 Mar 2023.

[8]The passenger pigeon has been used as an example, but there are plenty of others. Industrial, and even artisanal, fisheries kill animals that are caught in nets or hooks that are not the target of the fishery. These animals are killed and discarded. The threat may significantly impact the abundance of, for example, albatrosses, marine turtles, sharks, or dolphins. Likewise, bottom trawling is a modality of fishery that drags the seafloor, killing innumerable living creatures just as a by-product of the procedure. It will never be known if this fishing operation has extinguished entire forms of life.

11. IUCN Red List. Thylacine. *Thylacinus cynocephalus*. https://www.iucnredlist.org/species/21866/21949291. Accessed 11 Mar 2023.
12. Cafaro, P. 2015. Three ways to think about the sixth mass extinction. *Biological Conservation* 192: 387–393.
13. Foot, P. 2001. *Natural Goodness*. Oxford: Clarendon Press.
14. Wittgenstein, L. 1922. *Tractatus Logico-Philosophicus*. London: Kegan, P., Trench, Trubner.
15. Humphreys, R.K., and G.D. Ruxton. 2019. Adaptive suicide: is a kin-selected driver of fatal behaviours likely? *Biology Letters* 15: 20180823. https://doi.org/10.1098/rsbl.2018.0823.
16. Galimov, E.R., and D. Gems. 2021. Death happy: adaptive aging and its evolution by kin selection in organisms with colonial ecology. *Philosophical Transactions of the Royal Society*. https://doi.org/10.1098/rstb.2019.0730.

8 Represent-Evaluate

We can distinguish conservation practices according to the language-games they play in representing and evaluating living things. To represent, in the intended sense, is to describe what something is (Box 1.3)—its properties, what it does—and, as usual, our focus is on living things. To evaluate, in the intended sense, is to make a value judgment: a judgment about what is good or bad, permissible or impermissible, and many other more specific judgments, like that something is useful, beautiful, sacred. "Good," in the language of life forms, is primarily "natural goodness": which is constituted by a creature's satisfaction of the needs of its life form and, thus, what is required for its flourishing and autonomous life. Natural goodness is implicitly described by the propositions of natural history that represent what it is to be a creature of a certain kind and that, in describing the specific shape of its life, also at the same time implicitly lay down the standard for telling whether the creature is a good or bad representative of its kind (flourishing or dying, healthy or sick, sound or defective, etc.). We will develop these details of life form language in Chaps. 10 and 11. Natural historical propositions are then a representation or description of what a living thing *is* and also, at the same time, a standard for evaluating whether it is *good or bad*, in the sense of doing well or badly. Then, the language of life forms has the dual, interconnected power to 'represent-evaluate,' as we might say. Its representation and evaluation are two sides of the same coin.

Logical Consistency

Drake's explorers reported (Box 4.1):

> [We] were more and more overcharged with featherd enemies, whose cryes were terrible, and their poder and shott poisoned us even unto death, if the sooner wee had not retired and given them the field for the tyme...

C. Campagna, D. Guevara, *Speaking of Forms of Life*, Fascinating Life Sciences,
https://doi.org/10.1007/978-3-031-34534-0_8

The 'enemies' were possibly cormorants. "Feathered" represents them; "enemies" and "terrible cries" carry value judgments, not derived from the language of life forms. Finally, the interaction is evaluated as dangerous to humans: "they would have harmed us to death." This language makes all too predictable what follows in the story: "Having supplied ourselves, we returned the next day and many times more, to avenge ourselves on so barbarous an adversary and weaken his power." With this attitude, the men killed as many animals as they could.

The example illustrates a logical movement from representation, to evaluative judgement, to intervention. Drake's men identified and described the birds and their abundance, assessed them as bad in various ways, and on that basis found reason to intervene as they did, destroying the animals as the best way of redressing what they seemed to see as almost a matter of justice. The natural order for them had been disrupted; human hierarchy and dominance needed to be reestablished. The human act of 'interrupting procreation' of the animals was justified and 'natural.' 'Unnatural,' for them was the lack of control over the wild animals, due to the absence of humans in those parts of the world (the Tehuelche, or Patagones, culture, unlike Fuegians, was not marine). No doubt, the language used to represent and evaluate the passenger pigeon, before extinguishing it, reflected the same prejudices. The language of hunters would have probably suited Drake's description of birds colonies as "a stoare-house of victualls for a king's army." Also, a bird 'harmful,' 'evil,' and plentiful as a 'pest' can be but a 'feathered foe.' In the absence of another grammar of representation and evaluation, effectively describing the life forms for what they are, and the natural goodness they are capable of manifesting, the mass killing seemed obviously indicated and justified as the only logical conclusion.

Compare this to the following description of nature—critical too, but in a different way. Darwin reports, about what seems to be a species of caracara:

> These birds are very mischievous and inquisitive; they will pick up almost anything from the ground; a large black glazed hat was carried nearly a mile, as was a pair of the heavy balls used in catching cattle. Mr. Usborne experienced during the survey a more severe loss, in their stealing a small Kater's compass in a red morocco leather case, which was never recovered. These birds are, moreover, quarrelsome and very passionate; tearing up the grass with their bills from rage. They are not truly gregarious; they do not soar, and their flight is heavy and clumsy; on the ground they run extremely fast, very much like pheasants. They are noisy, uttering several harsh cries, one of which is like that of the English rook, hence the sealers always call them rooks... [1, p. 107]

Darwin's description is a bit more complex in terms of the language-games it makes use of: one involves the expression of qualities of temperament ('mischievous and inquisitive') and another involves aesthetic qualities ('their flight is heavy and ungainly'). The language-games vary and so, therefore, do the standards used to support the related value judgments. In the face of this language, the possible actions could be to continue to observe, and delve into the natural history, or to avoid the animal and prevent it from approaching objects important to the observer. That is, the differences in grammar for representing and evaluating facilitate a variety of distinct interventions.

This final excerpt is a natural historical description of the passenger pigeon and the goshawk, by ornithologist and artist John Audubon:

> Although the flight of our Passenger Pigeon is rapid and protracted almost beyond belief, aided as this bird is by rather long and sharp wings, as well as an elongated tail, and sustained by well regulated beats, that of the Goshawk. . . so very far surpasses it, that they can overtake it with as much ease as that with which the pike seizes the carp [2, p. 27].

This is what we are calling the language of life forms, where the standard of natural goodness predominates. It prompts a very different attitude of respect, wonder, and curiosity, and the inclination to continue one's observations—something that of course is no longer possible.

Primary and Secondary Values

We take the dual-function, 'represent-evaluate' grammar of natural history to be primary for conservation, in the sense that it should not be subordinated to other grammars, whatever secondary value they may have in describing or evaluating life forms. The natural goodness that the grammar of life forms implicitly represents cannot be subordinated to any other value if what we primarily care about is a flourishing, autonomous life (i.e., a life that does not depend on the satisfaction of the needs of another life form for its flourishing). One must evaluate according to the primary standard dictated by the form whenever a creature is declared good or defective, strong, fecund, healthy or sick, wounded or dying. The standard set by the form is what determines whether a bearer possesses or lacks essential qualities for living, or whether its behavior is as it should be, or whether it behaves strangely, etc. Conversely, the evaluative judgments that led to the killing of cormorants, passenger pigeons, thylacines—and many other species driven to extinction or near extinction—are based on the needs or interests of certain human beings, and are therefore not primary for the other life forms. In fact, none of the values behind the slaughter of these animals have anything to do with our natural goodness or theirs. Other animals can be frightening and annoying to us (as we to them!), but it is not therefore a necessity of our life form to slaughter them. The primary criteria of goodness or badness for a living thing are rooted in its form of life; criteria secondary to these often devalue or under-appreciate the individual, because they operate with standards distinct from and often alien to the form of life. On the standard of life forms, the 'barbarian adversary' was a *good* representative of its form.

Primary and secondary value judgments predispose us to different actions, and even when they may support the same action, they do so for reasons that are different. And, of course, the two types of value may support incompatible actions. Darwin, who was first and foremost a naturalist, was disposed to evaluate life according to the primary values implicit in natural history and it is hard to imagine him justifying or being part of a cormorant slaughter to restore order to the universe. Drake, presumably guided mostly by secondary values, would not have been suited to theorize about the origin of the species via natural selection. Though he might not call it such, a naturalist is poised to find the primary value of *natural goodness* in the behavior of creatures that sailors, trained to focus on other things, killed as enemies. An exquisite birdwatcher, such as Audubon, would have distinguished between a cormorant that builds a nest well and one that is less experienced, or between an individual with typical coloration and one that is feathered with a variant. These observations refer to the primary value implicitly expressed by the language of life forms, namely, the natural goodness of the animal, i.e., what it has and does to satisfy the necessities of its form.

Drake's men may have seen what natural history describes, but they did not prioritize its grammar in their representation and evaluation of nature, nor decide accordingly how they were going to intervene. The tendency of Drake's men, like so many others alienated from the primary standard for judging the well-being of living things, is to judge in terms of qualities that humans appreciate (cooperativeness with humans or 'intelligence') and to treat other life forms as of lesser value when they are thought to be lacking the favored qualities. The food industry, which exploits domesticated species, is an example of mistreatment based on a representation of the animals that devalues them to the point of treating them as though they were artifacts made for us, which indeed industrial farming has pretty much made them. Such animals are no longer thought to have qualities that can be reasons for respecting or protecting them. Conversely, if a non-human species is represented and evaluated, for example, as having 'rights,' or as a 'sentient' species, or as having 'intrinsic value,' it is made worthy even of 'moral' treatment, although the basis for this tends still to lie in values that are unrelated to the needs of the life form, i.e., to the primary values of the life form In short, the language we use to represent living things commonly favors qualities that prompt certain attitudes of approval and disapproval having little to do with natural goodness, and more with our homocentric projections onto them of what is valuable in a human being.

Evaluating Interventions

As we've been discussing, if a species is first a 'pest,' and only secondarily a passenger pigeon that migrates in large flocks, the reasonable intervention would be to eliminate it. Doing so is permissible and the eliminator is not required to justify his behavior further, or thought to deserve to suffer consequences for his actions.

Let us consider a more current example, one that raises some of the pressing issues of the day. 'Invasive' species are considered pests, or destructive species, and this is what is thought to justify their extermination, in the context of the conservation of a natural environment. But in the conservation of the forms of life the natural goodness of the living things has priority, and interventions (especially lethal ones) cannot be justified by appeal only to secondary values. Humans often introduce invasive species voluntarily, for hunting or fishing them, or for constructing certain landscapes, for example. Invasives are also a side effect of what is carried on boats, etc. If the invasive creatures are living according to the natural historical requirements of their form, and judgment is based on natural goodness, then on what grounds can we interfere with their lives? We may have to kill individuals that threaten our food resources or other vital needs. But in the language of ecology, the species is 'invasive' because it was introduced by humans into a native community whose ecological function is being profoundly affected by the introduced species. Is it then necessary for the satisfaction of the requirements of our form that we intervene on behalf of the affected 'natives?' Humans introduced the invasive, so do we bear responsibility for removing them, even if it means violating the life of the invasive, whose representatives may be actually satisfying the necessities of its form?

Our purpose here is not to decide all the many issues raised by invasive species, but to illustrate how, if operating with a standard of natural goodness, one would have to reason and struggle with the issues. In Chap. 26, we return to the question of the pressing issues of the day and how a virtue theory based on natural goodness would begin to treat them. For now, we wish to indicate how evaluating an intervention, like those commonly applied today, must consider the various language-games being played and relate them to the primary standard of natural goodness before deciding which interventions we are ultimately justified in taking or refraining from. In the case of invasives, the analysis would have to consider how our eliminating the natives is a necessity of our form (it is obviously not a necessity of the invasive's form that we eliminate it). What necessity would that be? We can attend to the constraints that the relevant grammar puts on an answer: in general, we cannot let secondary values settle the matter, as tends to happen because of our homocentrism. As we discuss in Chap. 26, ultimately the correct answers can only come from developing good judgment, based on the right virtues. The case of invasives is also complicated by the necessities of other forms, affected by the invasive.

Conservationists today are experts in language-games that appear to be a mixture of primary and secondary values. Conservation practices often justify interventions based on secondary values for life forms. We would maintain that it is only on secondary values that industrial whaling, trophy hunting, shark finning, wildlife trafficking, entanglement of dolphins in fishing nets, bycatch, and the clearcutting of forests and jungles are thought to be justified as permissible, and that those that carry out such actions are not judged as doing anything wrong or bad. These judgments are based on standards other than those of natural goodness, according to which the actions would presumably be *un*acceptable and *im*permissible.

As we will discuss in more detail later, judgments can be made about trophy hunting, and the rest, both as an evaluation of the activity and the character of the actors. On the standard of the conservation of life forms, the activities which affect the other forms of life must be judged according to how they affect their natural goodness, their ability to satisfy the needs of their lives. Trophy hunting, shark finning, etc., have a profoundly harmful effect on that, of course. And it is hard to see on the basis of natural goodness how either the activity or the actor could be justified. The language that supposedly justifies such things does so by, for example, representing the animals for their 'sporting' value, or of trophy hunting as an 'adventure,' or the hunter as a 'lucky adventurer' for having killed a vital, beautiful, wild animal. These language-games are the primary target for critical evaluation by the practice of conservation of life forms.

The language that each conservationist adopts is a large part of what determines the style and rationale of her practice. Depending on the language, provoked extinction may be understood as one of the gravest consequences of human behavior, and among the worst of the failures of conservation to stop the unacceptable, or, on the other hand, it may be seen as an inevitable consequence of meeting the needs of human development and economics. The language of representation and evaluation drives us toward one perspective or the other.

The Seeming Permissibility of Provoked Extinction Lies Implicitly in the Language of Representation

The language of Drake's sailors illustrates the attitude of superiority and entitlement that humans commonly assume they obviously have over nature. The attitude makes it reasonable to represent nature in way that makes its 'good' lie primarily in free access to our use of it and then, only after that, is there any concern for the lives of non-human species. The language we all commonly use is influenced by this attitude, even in conservation, and predisposes us to treat other species as a resource and to speak of them in the context of management and exploitation. When use values predominate, the practice of conservation sees itself as bound to correct whatever is detrimental to nature's instrumental value to us; that is the typical meaning of 'sustainability.' The way for an alternative, like the conservation of life forms, to displace the current dominant language and its logic is by installing its own logic and language and alternative system of evaluative judgments. Only then can the corrective interventions of the conservation of life forms begin.

It is imperative, then, for conservation to recognize that an essential part of its *practice* is to critically attend to the language of representation and evaluation that drives and justifies its interventions in nature. It must see this as in fact essential and central to the conservation of life forms. It must also make explicit the value judgments motivating and guiding its practice, and to avoid simply assuming them and leaving them unexpressed. As the errors in Wuhan, and countless comparable errors show, starting points matter. In particular, it matters how we describe nature, and living things in particular, in the first place. We must be wary of any interventions in nature based on representation of life forms other than that described by their natural history. The language and logic of natural history presents us with an intimately interconnected representation of life and value, one where every living thing carries the standard of its goodness in the life form it bears, the form that makes it a living thing in the first place. This is the only starting point, we contend, for investigating questions of why and how we should intervene on their behalf, or for grounds justifying whatever we do that affects them. It is the same grounds for judging our own goodness or badness in what we intentionally and routinely do to affect the other creatures for better or worse, and for judging what is lost, and what is wrong with us, in our staggering rate of annihilating the forms of life themselves. When we practice conservation outside of the logical structure of the language of life forms, the practice loses sight of the primary value of the flourishing of living things in all their diversity: the value represented by natural goodness. Any other way of representing life is bound to misrepresent the harm and loss of provoked extinctions. Indeed, it may justify actions fostering it.

References

1. Darwin, C. 1997. *The Voyage of the Beagle*. The Project Gutenberg eBook 944. https://www.gutenberg.org/ebooks/944. Accessed 11 Mar 2023.
2. Audubon, J.J. 1840. *The Birds of America*, Vol. 1. New York: J.J. Audubon, Philadelphia: J.B. Chevalier. https://books.google.com.ar/books/about/The_Birds_of_America.html?id=x1MDAAAAYAAJ&redir_esc=y. Accessed 11 Mar 2023.

Standards

9

This chapter discusses the concept of a standard for evaluative language. Our discussion is somewhat general, but our main concern is, as always, the evaluation of provoked extinctions. In general, standards matter because, among other things, of the way they can guide and sustain a value system, and because of their role and limits in resolving conflicting evaluative judgments. In the language of life forms, the standard of goodness is a creature's autonomous thriving in its natural habitat. Creatures who meet the standard live according to the dictates of their form, and therein lies goodness according to this standard, primary for life form conservation. The standards of goodness that guide interventions, as expressed in current conservation discourse, are typically secondary to that of an autonomous and flourishing life, as is the case with aesthetic language ("beautiful," etc.) and instrumentalist language ("useful," etc.), with their varying standards relative to tastes and interests. Only the language of life forms contains the standard of natural goodness, a standard based on the logically interdependent relation of life form to bearer, and thus universally and objectively the same for all life.

In ordinary language, standards of evaluation are generally understood from the context and not usually made explicit. Language users do not usually have to clarify the standard for the judgments they express. Among Drake's seafarers, for example, there was no need to explicitly justify the act of killing each animal by expressing the standard, although it was suggested that 'good' was man dominating nature. Generally, our ordinary use of language allows us to express ourselves, and to communicate with and understand each other, effectively and naturally, through context, shared assumptions, and our ability to interpret each other. Of course, confusion does happen, and perhaps even some 'bewitchment,' in ordinary language use, but as we go about our daily business these are not the norm, the way they are in philosophical contexts. Inasmuch as all conservation practice is based on a philosophy (knowingly or not), making the standards of evaluation explicit is very important, especially when assessing a state of affairs that may lead to the extinction of a species. As we have been emphasizing, the standard greatly affects the nature and

C. Campagna, D. Guevara, *Speaking of Forms of Life*, Fascinating Life Sciences,
https://doi.org/10.1007/978-3-031-34534-0_9

effectiveness of the practice, and its justification and motivation, or, in a word, its philosophy.

Now, making explicit a standard does not always lead to needed reassessment and revision of it. Making explicit the standards underlying what was done in the Wuhan market, for example, would not have prevented the mistakes made there, because the implicit standards were uncritically accepted as fine, even obvious, as with Drake's men. However, we might expect it to be different in the community of conservation practitioners, at least where the reason for being is to identify and eliminate threats leading to provoked extinction, out of concern for the life forms themselves. Considering the variety of standards now in play in the various language-games in conservation, it is necessary to be explicit about them, as we will discuss more fully in Chap. 23.

We can divide the standards that guide conservation practices into two categories: those that guide evaluative judgments about living things (and the rest of nature) and those guiding evaluations of human actions that affect them. For example, an aesthetic standard of goodness is often appealed to in conservation when judging a creature's value, e.g., beauty. But, of course, as we have been emphasizing, economic standards are also pervasive: 'resources' are good (their lack, bad), and these are understood in terms of the 'natural capital' that living nature provides, all of which are 'useful to economic development,' etc. We have noted that in the framework of population biology, 'good' is a stable or growing population. Within this framework, human actions like trophy hunting may be deemed permissible, even desirable, if they contribute funding to a park at no risk for the size of the population, for example.

Regarding evaluations guiding human action, our focus is on the human acts which affect living things, and on very general rules or principles that might guide our actions, like that it is permissible to kill for food. "Human beings kill in order to obtain food" expresses a natural historical judgment and thus a natural necessity of our form, an instance of natural goodness. And it is from the system of such judgments that we hope to derive some general principles or ideals. So far in our exposition, "good" and "permissible" are being used in a very general way, as they are here in the general claim that it is permissible for us to kill to eat. More specific questions of *what* we can kill to eat, or *when* and *how much*, must be worked out as theory and practice develop. At first the rather impoverished terms "permissible" and "good" are just familiar points of entry into an extensive and richer investigation of questions of value that involve a variety of language-games, with their glossaries and grammars. At first, we seek general principles; after that we must learn to apply and develop them in our effort to fully express values as yet uncaptured by language.

Primary and Secondary Goodness

If a whale is assessed as good and, for that reason, shot straightaway with a harpoon, then "good" is being used according to standards alien, and thus *secondary*, as we are calling them, to those implicit in the life form; it is being applied to what is useful for whaling purposes (e.g., size or market value), rather than what pertains to the satisfaction of life form needs. On this standard, the animal is still 'good' even when wounded or dead. The language goes so far as to introduce names, like "Right Whale," applied to two species of mysticetes of the genus *Eubalaena*—so called because they are slow and float once dead and have, therefore, the 'right' qualities from the standpoint of whale hunters! So far as the language-game of commercial use has operated for centuries, there is no need to be concerned with judgments of natural defect, cruelty or violence, except as they may affect economic efficiency, in obvious contrast with life form language. On yet other standards, e.g., traditional aesthetic standards, the harpooned whale might evoke judgments of disgust and repulsion, or on the other hand, compassion and anger, of the kind expressed by the *Save the Whales* movement, where the standard is well-being.

When evaluative language diverges from what we are calling the *primary* standards, i.e., those set by the needs of life forms, and instead are based on standards at best secondary to those needs, the two language-games for evaluation may conflict and give rise to disputes, cost-benefit considerations, or pluralistic solutions. Secondary standards for the language of life forms may originate, for example, in economic or social theories, traditions, feelings, aesthetic appreciation. If what we are concerned with is the flourishing of living things, then these standards lack the generality and objectivity of the standard behind natural goodness. Moreover, there is no evident common basis for resolution of conflicts, no single overarching standard, like that guiding judgments of natural goodness. For whalers, the scene of a dead whale floating in a bloody sea may not be an unattractive sight. Standards of beauty vary with cultural and individual tastes. Similarly, spiritual standards, as much as those of use value, vary according to a wide variety of standards relative to culture, opinion, and faith.

The standard implicit in the language of life forms applies universally to all living things, inasmuch as it is found in description of the form which makes them living things in the first place. The standard has the same objectivity as any natural historical proposition in a monograph which describes a species, for those propositions are in themselves the standard; they express the criteria or conditions which determines whether any individual representative of the species is alive and well, or not.

Now, there is some language in conservation that means also to express objective and universal values: e.g., the language of 'intrinsic value,' 'rights,' or 'animal welfare.' For example, in the animal welfare movement, sentience is considered universally and objectively valuable in so far as anything that can feel pleasure or pain must be treated with the same ethical respect (Chap. 14). On this view, 'animal rights,' like human rights, are thought to be based on a broad, objective feature of the world, in this case, sentience (rather than, say, autonomy or intelligence, which are more limited in their extent). Pleasure and pain are intrinsically and objectively good and bad, respectively. Still, even these standards are not universally and objectively relevant to all life forms. And so they are questionable as a standard for anyone primarily interested in trying to express the 'intrinsic' or objective and universal value in living things. Even when the various evaluative judgments happen to coincide, the various standards give different reasons for intervening in nature and, in particular, for redressing provoked extinctions. This pluralistic scenario, reflecting multiple standards, expressed by various language-games, is often seen as desirable despite the problems it poses of commensurability, among others (Chap. 16).

In contrast, we have in the language of life forms, one general, natural, and objective standard of well-being for all living things, as natural and objective as the biology of a jellyfish or any other life form. Any disagreement about whether this should be the primary standard for conservation of all living things will need to present alternatives. But if "conservation" means "the conservation of the well-being of living things, in accordance with all their diverse forms," then it is hard to see a better starting point, or any need to compromise or negotiate with standards conflicting with natural goodness. It is a sad irony that, especially in conservation biology, judgments based on other standards are given priority over natural goodness (if natural goodness is even explicitly recognized at all). In the conservation of life forms, the only justification for our interfering with the other life forms is for the sake of the satisfaction of our own life form needs. And we must develop the wisdom and good judgment not to mistake what we simply prefer, or are used to, with what is a true human need, as expressed by the relevant natual-historical necessitites describing our form.

Conclusion

In Indonesia, birds may be captured and traded to compete in singing competitions [1]. These competitions are associated with economic benefits. In Cyprus, birds are trapped with glue, killed, and then used in the preparation of traditional foods [2]. There are many other cultures that accept the practice of trapping birds, even small ones, to sustain culinary traditions. In other parts of the world, on the other hand, the preference is to observe birds in their natural environment and for pleasure. As a result, birds are fed, nesting opportunities are provided, the environments on which they depend are protected, and trapping them for food is not permissible as a social practice. This is done for the good of the birds and of the humans that appreciate them aesthetically, among other benefits. These and countless other examples illustrate the variety of standards that guide or constrain our actions with regard to other living things.

In the absence of an objective and universal and relevant standard, like natural goodness, then concepts like sustainable development succeed and continue to reign, in part because they offer a style of evaluative discourse that is convenient to the human way of life. It does not matter that for the language of sustainable development to be possible in the first place, the representative-evaluative language of the natural goodness is needed, since otherwise we cannot speak of 'sustaining' *natural life* at all. The essential language for the representation-evaluation of life forms has always been at the disposal of conservationists, but many prioritize alternatives, placing the natural historical component under the umbrella of the science of biology, for example, or shifting to aesthetic or instrumental standards.

Today, a practice based exclusively on primary standard of the form of life is at best rare, and generally considered unrealistic and ineffectual where it might be said to exist at all.[1] There are conservationists who may avow this standard as the underlying reason for their practices, but they actually use other standards when making their evaluations explicit. We should not be surprised by what happens: a mismatch between the explicit value judgments about what should be done to redress threats to nature and the needs of the forms of life. The more those needs are subordinated to political judgments of sustainable development, the more the interventions of conservation will be pulled this way and that, relative to unstable and even volatile forces, alienated from the natural goodness of the forms.

Pragmatists who would say that the standard of natural goodness is unrealistic, and not likely to bring about the intended results, will ask questions like: Why should a fishery revise its priorities based on the natural needs of target species to aggregate for spawning? Why should rain forests be preserved just because of the great variety of insects that depend on the existence of the canopy of trees, worth a lot as lumber? But, of course, it is highly questionable whether the pragmatic attitude has itself delivered good results, given the appalling and unprecedented rate of human-induced extinction. What we do, and why we do it, is as important as any result (often beyond our control). The *means* are the ends. In any case, we contend that only practices based on the conservation of life forms should be considered any longer, and all other values subordinated to the primary standard for life forms. It is a safe prediction that continuing with the practices as they are now, provoked extinctions will continue apace, and unless standards are made explicit, it will never be clear how they conflict with each other, or how some secondary values may be confused for primary ones, or how the confusion among standards may undermine the effort to eliminate the threat we pose to species, or obscure the primary reasons why we should be concerned with their conservation in the first place.

[1] The *Save the Whales* movement is a compelling counter-example to this view, inasmuch as it was at least for a time extremely effective and based partially on the primary standard of natural goodness [3]. See Chap. 20, and Boxes 20.1 and 20.2.

©Andrew Guevara

Box 9.1: Non-transferable Human Standards

In the effort to protect them, certain environmentalist practices argue for rights for animals, based on animal intelligence or sentience or other favored similarities to humans (see Box 14.1). These practices transfer human standards to other forms, and justify doing so with language that resembles natural history expressions, even though the grammars differ. For example, "orcas have a right to life because they are intelligent animals" and "orcas

(continued)

Box 9.1 (continued)
learn behaviors that they pass on to their offspring" are taken on par in their logical grammar. The latter is a fairly uncontroversial natural historical judgment, with the grammar of an Aristotelian categorical. The concepts of learning and behavior are fairly clear and stable in ethology and evolution; the language-games are fairly well agreed upon. Talk of "rights" or "intelligence" is not uncontroversially the same in this respect. The reason for this is not, however, that orcas are not intelligent or, even, that they could not have rights (in some sense) based on their intelligence. The issue is with the language-games of "intelligence" and "rights." There are so many ways of being 'intelligent,' from robots to ants. Not all intelligence is necessarily of the human form. The term is, in this way, fraught, like "sentience" or "consciousness" and what it is like to be sentient or conscious, etc. (see Chap. 14). Likewise, any talk of rights based on intelligence will have to orient itself to the many language-games before assuming that it can, or how it can, go into a legitimate categorical that describes the non-human life form, without turning it into a human being that can, for example, enter into a social contract, and make promises and use language like we do, and so on.

It is for such reasons that the practice of life form conservation, as we will discuss in Box 14.1, does not treat rights claims, or intelligence, as a primary basis for conservation of other species. The uncritical application of standards across life forms often leads to evaluative errors that exacerbate the threat to species. For example, it is common to devalue many living things, by judging them to be lacking in favored human attributes, such as intelligence (Chap. 24). Human 'superiority' has generally been sustained by the anthropocentric view that non-human living forms are 'imperfect' by our standards. The language of life forms does not allow for such confusions. The standards of evaluation are non-transferable but necessarily relative to the form of life.

References

1. Paddock, R.C. 2020. Bought for a song: An Indonesian craze puts wild birds at risk. *The New York Times*, April 18.
2. Franzen, J. 2010. Emptying the skies. *The New Yorker*, July 19. https://www.newyorker.com/magazine/2010/07/26/emptying-the-skies. Accessed 11 Mar 2023.
3. Campagna, C., and D. Guevara. 2022. "Save the Whales" for Their Natural Goodness. In *Marine Mammals: The Evolving Human Factor*, ed. G. Notarbartolo di Sciara and B. Würsig, 397–424. Cham: Springer. https://doi.org/10.1007/978-3-030-98100-6_13.

The Unique Logic of Life

10

What merely 'ought to be' in the individual we may say really 'is' in its form.
M. Thompson [1, p. 81]

What is a form of life? About this, Thompson says:

> [What] I have called a life-form… I have also called… a species, with some reservations, and would be happy, in an Aristotelian mood, to call it *psuche*. But each of these latter expressions carries a baggage of associated imagery—a picture to hold us captive, if you like… Or, perhaps, if we stress the "form" in our preferred expression "life form," the thing will even be sought in a Platonic heaven or in the mind of God… But all such images should be cast aside. I think our question should not be: What is a life-form, a species, a *psuche*?, but: How is such a thing described? [1, p. 62]

It is also helpful to recall our earlier citation from Wittgenstein regarding the definition of terms: "Let the use *teach* you the meaning" [2, §250]. The phrase "form of life" (or "life form") is already in use in this work and we hope to have been illuminating its meaning thereby. The answer to our opening question lies, then, in the grammar of how the thing is described. This chapter examines closely the logical structure of the language in which the notion of a form of life originates. We owe the logico-grammatical examination of this language to the philosopher Michael Thompson [1]. We will be here only summarizing his foundational work.

Like Foot, Thompson has *not* applied his results to the disciplines of conservation; that application is part of our own contribution, together with our application of Foot's theory of natural goodness, also based on Thompson's analysis. Ours is but one potential development of their rich and path-breaking theories. In this chapter, we concentrate on Thompson's logico-grammatical analysis of the language we think should guide the practice of conservation of life forms: a practice that has its reason for being in the preservation of natural goodness—the autonomous satisfaction of a creature's life form needs.

C. Campagna, D. Guevara, *Speaking of Forms of Life*, Fascinating Life Sciences,
https://doi.org/10.1007/978-3-031-34534-0_10

The Logic of Natural Historical Judgments

Thompson demonstrates that it is necessary for us to be able to represent living things according to a certain logical form, without which it would not be possible for us to so much as experience or identify them among the objects of reality. As he notes "... an intellect cannot have a power of apprehending objects unless it has a power of thinking something *of* them" [1, p. 57]. He shows that to think something (anything) of a living thing requires that we understand, and apply to our experience, however roughly, the representation of a certain logical schema, or template, that allows us to apprehend the shape of the creature's life. This point is critical for his view, as for ours. It means that all our judgments about living things, and the language we use to express them, involve a representation of this logical form.

In Chap. 6, we cited Ursula Le Guin's story, whose character, disenchanted with language, "unnames" all living things; the act of abandoning names is symbolic of her dissatisfaction, as it is not names that matter the most. As Wittgenstein says [3, §49]:

> For naming and describing do not stand on the same level: naming is a preparation for describing. Naming is not yet a move in a language-game — any more than putting a piece in its place on the board is a move in chess. One may say: with the mere naming of a thing, nothing has yet been done.

Identifying a form of life is not in the naming of it. The language requires further moves, like the description of doings of the 'vital' sort: eating, breathing, blossoming, mating, reproducing, or the description of certain attributes such as bones, shells, flowers, feathers, and hair. The notion of an organism is formed by describing what the thing is, has, or does, in terms of the shape of its life—a language grounded in a unique logic. The focus here is on the logic behind the specific language and terms, and with a helpful presentation for the non-expert, of a rather technical subject matter.

Logic, as Thompson notes, "relates to the *form* of thought—a form of inference, for example, or the 'logical form' of a judgment. 'Form' here is of course opposed to '*content*'" [1, p. 26]. In a traditional, elementary logic course, form is distinguished from content by the use of schematic letters. For example:

Fa,

where the capital letter '*F*' stands for a predicate and little '*a*' for the subject of a declarative sentence. *Fa* can represent, for example, the sentence "The cat is on the mat," where $F =$ *is on the mat* and $a =$ *the cat*. The content fills in the blanks of the form. More complex sentences can be represented this way:

Fab,

where, e.g., *F = is the father of.* And we can generate more complexity still, with *F = is the father of [blank] and [blank].* The logic course teaches techniques for dealing with even more complex structures, built up from these simple schemas. In traditional grammar, we are taught to identify the subject of the sentence, in simple declarative sentences like "The cat is on the mat." When things get more complex we are taught to distinguish between 'subject' and 'object', as in "Daniel is the father of Izcalli." But, of course, we can get the same thought across with "Izcalli is the daughter of Daniel". So what's the subject and what's the object? Apparently, it can be reversed at will! However, all that really matters for the logical form is that we understand the distinction between predicates, with their place holders, and the content that fills them in.

This way of representing the form of a declarative sentence (i.e., a sentence that can be true or false) is due to Gottlob Frege, the founder of modern logic. It is a simple but very powerful formalism because of the way it allows us to display fundamental logical relations and the logical inferences based on them. The most fundamental relations are those expressed by negation, conjunction, disjunction, and conditional, among others. For example:

Not Fa
Fa and Fb
Fa or Fb
If Fa then Fb

With these, we can now describe elementary logical relationships: e.g., *Not Fa* is true if and only if *Fa* is false. The conjunction *Fa and Fb* implies *Fa*. Of course, it also implies *Fb*. The conditional *If Fa then Fb*, together with the antecedent, *Fa*, imply *Fb*.

A first course in logic shows how we can derive, in this way, quite a powerful range of rules and theorems, useful for formalizing the logic, for example, of important parts of mathematics (Frege's original concern), and much more powerful and transparent than the traditional syllogistic logic going back to Aristotle.[1] The reader familiar with elementary logic will likely associate logic with the rules of inference and proof procedures for deriving theorems. But to appreciate Thompson's contribution to the logic of life forms we must attend to the idea of *the logical form* of a sentence and corresponding thought.

Frege's simple and elegant notation, *Fa*, etc., means to represent two basic categories of thought, corresponding to the predicate of the sentence, and what fills it in to complete the sentence, or, in other words, to a property and the object to which the property is being attributed. This is a super-abstract sense of "property" and "object"; it does not matter whether we are talking about numbers, God, characters in a novel, atoms, cells, species, etc., and whatever properties these may

[1] Mainly due to the introduction of quantifiers; see [4].

have. This is in keeping with our interest in form, rather than content. The purpose of the abstraction is to represent the fundamental structure of any thought that has the potential to apprehend reality: i.e., to think something of an object. Then, the logical forms of sentences and thoughts are the minimally necessary structures for apprehending or expressing any content of thought.

What Thompson shows is that our apprehension of living things requires a logically unique form of thought. One needs some understanding of the logical technicalities to fully see why this is so, but the essential insight can be appreciated by comparing two types of generalization, e.g.:

All horses have four legs

and

Horses have four legs
(or equivalently*: The horse has four legs*)

If, as is typical, what we mean by the first generalization is that every horse has exactly four legs, then the logic of this generalization—called "universal generalization"—has the following implication: the generalization is false if there is at least one horse with less (or more!) than four legs. The second generalization is an example of what linguists call a "generic," because it refers to a *kind* as opposed to a *particular instance* of a kind [5, 6]. Some generics can be stated as universal generalizations or statistical generalizations, generalizations over individuals in a population. But this is not so of what Thompson calls *natural historical judgments.*

When the generic expresses a natural historical judgment, as we are assuming it does in "Horses have four legs," it says something about the *form of life*, *horse*, as opposed to describing any *individual* horse or horses as members of a population. So here, unlike "The cat is on the mat," the subject of the sentence is not an individual animal, but a kind of animal and this changes the nature of the implications. When understood as a fact about the kind, or form, of life, we do not retract "The horse has four legs," even if most individual horses, through injury or disease, tragically come to have less than four. The point is perhaps clearest when we consider that individual creatures of any kind largely fail to make it through their developmental cycle, being picked off by predators, or just failing in some other way due to external conditions, or internal defect. Most acorns do not become oaks.

Then, thoughts about life forms cannot be represented by universal generalizations; the latter have the wrong logical form, as revealed by their implications. The reader may recognize application of this point already, in our critique of population language, whose logic is that of universal or statistical generalization: generalizations over individuals. But we are here isolating and focusing on the logical form of life form language and judgment, and showing how it is unique and rock bottom. It cannot be represented with the logical form of any other type of generalization, nor universal generalization, nor statistical, nor with the use of 'all things considered' clauses, or clauses about 'normal conditions.' For example,

All acorns become oaks under normal conditions (or all things equal)

is unhelpful as a paraphrase of the natural historical judgment, since what is 'normal' for the kind can only be determined by the shape of its life [1, pp. 70–73].
The life form generic *does* have, however, 'normative' implications for individual bearers of the kind: e.g., it is why we expect individual horses to have four legs, and why, if they do not, we think of it as due to a defect or injury, or other accident (natural badness). The representation of the form of life implies what 'ought' to be in the individual that bears it, that is the normative implication.

'Aristotelian Categoricals'

The previous section sketches the unique logic of a peculiar form of judgment that, Thompson shows, is necessary for apprehending living things. Again, some technical knowledge is required to fully appreciate this point, but it can be seen, partly, in the logical difference between generalizations over particular individuals, and the general claims about *the horse* or *horses*, etc., when what is being referred to is a form of life. This chapter started by asking what a form of life is, and with Thompson's observation that the answer lies in how the thing is described. He summarizes the results of that description as follows:

> A species or life-form is. . . the sort of thing to be the subject of a general judgment or a general statement; it is the sort of thing that is said of something and about which something can be said, in the sense of Aristotle's Categories [1, p. 48].

The general judgments here referred to are typical of those made by naturalists when describing the creatures that they discover and study: i.e., *natural historical judgments*, as we are calling them, following Thompson. And, as we will see, the domain of objects they describe is "categorically" different from any other domain, in a sense of "category" going back to Aristotle, which is why we have adopted Thompson's talk of 'Aristotelian categoricals.'

Here is an example from Charles Darwin:

> . . . flying squirrels have their limbs and even the base of the tail united by a broad expanse of skin. . . [7, p. 165]

In this statement, Darwin is not generalizing over individuals: "The flying squirrel has limbs, etc." shares the logic of "The horse has four legs." Darwin is generalizing over a life form; he would not take the generalization to be necessarily disproved by observations of flying squirrels lacking a broad expanse of skin uniting the limbs or tail. If in fact a large number of squirrels, in a particular place or time, were found to lack the anatomical feature, the naturalist must determine whether this is a widespread defect, before revising his monograph. As Thompson notes,

> A natural-historical judgment may be true though individuals falling under both the subject and predicate concepts are as rare as one likes, statistically speaking [1, p. 68].

Thompsons's observations about the logical uniqueness of the representation of life run contrary to the near universal tendency to think that judgments about life express purely empirical facts about individuals, and generalizations based on those facts. But, on the contrary, if they are facts about life, they presuppose the a priori logical form that Thompson has described, with the logical grammar of the representation of life.

The discipline of natural history is full of statements identical, in the underlying logic, to Darwin's descriptions of the flying squirrel. Aristotle, who pioneered natural historical descriptions, had some grasp of the particular logic that applied to the representation of living things. He also developed the first system of categories, enumerating "the most general kinds into which entities in the world divide" [8], such as place, relation, substance, quantity, and quality. The idea is that the concepts of place, relation, quantity, etc., are so fundamental that they cannot be reduced to or explained according to some other concept, and that they are, furthermore, necessary for the apprehension of a certain domain of reality: physical things, for example, or numbers.

When Thompson then declares, as in the citation above, that forms of life are things about which something can be said, in the sense of "Aristotle's Categories," he is asserting that the concept of life, of living things, is its own fundamental concept or category, irreducible and necessary. Again, as Thompson says,

> It is because in the end we will have to do with a special form of judgment, a distinct mode of joining subject and predicate in thought or speech, that I am emboldened to say that the vital categories are logical categories [1, p. 48].

Thompson's exposition of the concept of life as a fundamental category is rigorously based on his logico-grammatical exposition of the Aristotelian categoricals that describe a form of life. He introduces his own notation (slightly different from the traditional Fregean *Fa*) of *S is, has, or does F*, as his preferred way of schematizing the basic form of Aristotelian categoricals, where "S" (the subject) stands in for one or another form of life, and "F" for the predicates that apply to the form. Where S = flying squirrel we can get:

The flying squirrel is a rodent mammal,

where F = rodent mammal.

This notation can also be expressed in standard, Fregean logical notation, but that is a technicality that need not concern us here. The point is that these and many other propositions of the same general form will fill the monograph on the relevant species. As, for example,

> The flying squirrel *is* a rodent that *has* its limbs and the base of its tail united by an expanse of skin, and that *glides* from tree to tree...

which describes what the form is, has, and does.

When individual flying squirrels—empirically identifiable, particular things here and now—are represented generically, through the logic of the forms of life, the subject of representation is no longer the individual squirrel, or population of individuals, but rather the kind or form of life *S*. Thus, as Thompson notes [1, p. 69], there is a deep syntactic ambiguity in sentences like: "*The squirrel jumps from tree to tree.*" The determinate meaning depends on whether the subject refers to the form of life—flying squirrel—as opposed to a particular bearer of the form: e.g., our pet squirrel, Rocky, that someone was asking us about, wondering what it was doing up in that tree. In the one case, we mean to assert a necessary truth, but not in the other. This latter meaning is like "The cat on the mat," a statement about a particular animal and not a natural historical judgment about its kind. As Thompson states:

> It is very easy, in large generalizations about "life" and "organisms," to overlook the possibility that one's propositions have this kind of generality, instances of which will themselves be a kind of generality, and not facts about individual living beings [1, p. 67].

A Unique Teleology

Thompson's exposition of the categorical representation of life is, in all its details, a topic that we will develop as we go. But there is one logical feature of Aristotelian categoricals that merits immediate consideration; it has to do with the unique *teleology* of their grammar. We will further discuss teleology in Chap. 12 and, especially, Box 12.3. For present purposes, we can continue to rely on the example from Darwin:

> ... flying squirrels have their limbs and even the base of the tail united by a broad expanse of skin, which serves as a parachute and allows them to glide through the air to an astonishing distance from tree to tree [7, p. 165].

We find similar statements in field guides and scripts of a natural history documentaries, and other familiar places. In this description, Darwin starts with the anatomy but follows immediately with the role that the expanse of skin plays in the life of squirrels: it serves as a parachute, etc. The thought is teleological or purposive: the expanse is there in order for the squirrel to glide from tree to tree. A naturalist's monograph is filled with such teleological connections (implicitly or explicitly), in its description of the life form. The teleology is not a metaphor or mere heuristic; it does not describe accidental or coincidental relationships. On the contrary, one cannot understand what this organism is, has, and does without the teleology. So it is with all organisms. Moreover, this is a unique teleology: it cannot be identified with the teleology of conscious purpose or design. The logic is different, as we will further develop in Chap. 15 and Box 12.3.

The teleology of Aristotelian categoricals has the logic of *conjunction*, while the teleology of intentional purpose or design does not. The teleological connections in "The oak sends down tap roots, deep into the soil, in order to obtain water and nutrients, in order to grow and develop bark, leaves and fruits. . ." imply "The oak sends down tap roots *and* the roots obtain water *and* grow and develop bark, etc."; i.e., if the purposive judgment is true so is the conjunction. But this is not so of: "The Red List is published in order to inform us of the threatened status of many species, in order to slow the rate of provoked extinction." This does *not* imply the conjunction: "The Red List is published *and* we are informed of the status of species, *and* the rate of provoked extinctions is slowed." The fact that there is a Red List does not mean that the rate of provoked extinctions is slowed; in fact, sadly, we know it is not. The teleology of the representation of life might at first glance be thought to require an intelligent designer of the universe, contrary to our modern scientific understanding as not requiring one. So, a clarification of its distinctive grammar removes this potential stumbling block.

The 'Definitions' of Life

Biology presupposes the category *form of life*, as any study of life must. However, there is hardly ever any explicit awareness or primary concern with the categorical understanding of life and the logical grammar of life form propositions. Instead, as Thompson illustrates, biology attempts to 'define' life through 'characteristic signs' of life common to individuals. A typical list will include "reproduces" "maintains homeostasis," or "has DNA." But these, and other typical members of the list, as Thompson argues, are not sufficient (not even jointly) for differentiating the living from the non-living. Thompson notes that what is lacking is what philosopher Elizabeth Anscombe has identified [1, pp. 53, 54] as the 'wider context' necessary for the identification of particular living things. As Anscombe says,

> When we call something an acorn, we look to a wider context than can be seen in the acorn itself. Oaks come from acorns, acorns come from oaks; an acorn is thus as such generative (of an oak) whether or not it does generate an oak.

And Thompson, making essentially the same point, says:

> When we call something eating, then, we appeal to something more than is available in the mere spectacle of the thing here and now [1, p. 54].

That *something more* that makes it possible to understand that this thing here and now is eating is the comprehension of the form of life.

There is no substitute for considering all the compelling arguments themselves that Thompson gives to demonstrate this point about the inadequacy of typical attempts to define or adequately characterize life. We cannot summarize them all here but we urge the reader to consider them carefully [1, p. 33]. The basic lesson is always the same. The error lies in the attempt to characterize or define life through common properties shared by particular individual things, identifiable here and now, without any thought of or reference to a wider context. Without that necessary 'turn of thought,' without the language that goes beyond the thing here and now—e.g., to the developmental cycle or stages in the shape of a life—it would not be possible to experience anything as eating, or photosynthesizing, or hibernating, or as manifesting any other vital function or property.

Perhaps the example that makes the point best regarding the necessity of the form of life concept is mitosis. In an amoeba mitosis is part of the reproduction of the individual; in a human being it is part of the growth or self-maintenance of the individual (reproduction is another matter, involving other matter, as Thompson puns). What mitosis *is* necessarily depends on a 'wider context' than can be found in what is observable here and now. As Thompson says [1, p. 55]:

> The distinction between the two cases of mitosis is not to be discovered by a more careful scrutiny of the particular cells at issue—any more than, as Frege said, the closest chemical and microscopic investigation of certain ink markings will teach us whether the arithmetical formulae they realize are true.

No language-game that represents life simply in terms of characteristic marks leads to an understanding of life, without already relying on the grammar of life forms. In other words, not just any kind of true description will do. Therefore, and we will return to this point many times, a practice of conservation based on the logic of representation of life forms does not require a new vocabulary, a new definition of the nature of life, but, rather, a redirection of language back to the correct grammar, which we master from a fairly early age in ordinary contexts, and that can be expressed canonically by Aristotelian categoricals. These implicitly express the standard of natural goodness and, as we argue, the reason for conserving life in the first place.

Concluding Remarks

In the following quotation, Wittgenstein contrasts the strangeness of our trying to apply the qualities of the living to a stone, with the ease with which we apply them to a fly.

> Look at a stone and imagine it having sensations. One says to oneself: How could one so much as get the idea of ascribing a sensation to a thing? One might as well ascribe it to a number! And now look at a wriggling fly, and at once these difficulties vanish, and pain seems able to get a foothold here... [3, §284]

We here use Wittgenstein's words to stress the peculiarities of living things, made evident by a language so familiar that Thompson is prompted to say:

> It may seem a bit absurd that a form of predication suggestive of field guides, dusty compendia, and nature programs should be supposed to be the ticket for a philosophy of organism [1, pp. 66–67].

But the fact is that humans apprehend life by following a systematic, precise, interconnected, and unique logical grammar. Our main purpose in attending to the details of this logic, this grammar, is to show its power and depth as a foundation for clarifying the foundations of conservation practice, and for guiding its representation and evaluation of that which it means to conserve.

©Andrew Guevara

Box 10.1 The Language of Aristotle, Darwin, and Other Naturalists

What we have cited from Darwin about flying squirrels [7, p. 165] serves as a paradigmatic example of the natural historical judgments that represent life forms. Other examples from the same naturalist are:

> The [bee of the genus] Melipona... forms a nearly regular waxen comb of cylindrical cells, in which the young are hatched, and, in addition, some large cells of wax for holding honey [7, p. 205].
>
> In Patagonia, the condors, either by pairs or many together, both sleep and breed on the same overhanging ledges. In Chile, however, during the greater part of the year, they haunt the lower country, near the shores of the Pacific, and at night several roost in one tree, but in the early part of summer they retire to the most inaccessible parts of the inner Cordillera, there to breed in peace [9, p. 3].
>
> The Guanacoes have one singular habit, the motive of which is to me quite inexplicable, namely, that on successive days they drop their dung on one defined heap [10, p. 28].

The judgments expressed here, notwithstanding the metaphorical language, have the logical grammar of what Thompson has called "Aristotelian categoricals," after Aristotle who pioneered natural history and understood something of its distinctive features. Thompson notes:

> When Aristotle says that some animals are viviparous, he does not give Helen and Penelope as *examples*; his examples are: *man, the horse, the camel* [11]. His thought may thus be canonically expressed as "For some terrestrial lifeforms S, the S is viviparous." And when he says that some animals shed their front teeth, but there is no instance of an animal that loses its molars, he will not give up the sentence when faced with a denture wearer; denture wearers aren't the 'animals' he was talking about [1, p. 66].

Aristotle understood that in making generalizations that describe life, we make reference in the first place to the form of life, and not the individual bearers. Elizabeth Anscombe, Peter Geach, and a few others were among the philosophers in our own time who alerted us to the unique logic of these kinds of propositions (e.g., [12, 13]). Anscombe's own example was "Humans have 32 teeth," a requirement of the life form, even if most individuals are defective in this respect. The grammar of such propositions and their categorical nature is explained in more detail in Chaps. 10 and 12, and in Boxes 10.1, 10.2, 12.1, and 12.3.

Darwin's descriptions add color to the pure natural historical information. He begins with the natural history of the condor's form of life when he says that in summer they undertake migrations to remote places, but when he speaks of them reproducing "in peace," he switches to a language-game of metaphor or analogy. Still, even here he probably has in mind something that could be expressed as an Aristotelian categorical: a life form necessity about

(continued)

Box 10.1 (continued)
the kind of habitat needed for reproduction, for example, one relatively free from predators.

However, in the following citation, as in the others just noted, descriptions of natural history are mixed up with other language-games, or expressed through metaphor or colorful language:

> The condors may oftentimes be seen at a great height, soaring over a certain spot in the most graceful circles. On some occasions I am sure that they do this only for pleasure, but on others... they are watching a dying animal, or the puma devouring its prey [14, p. 319].

What Darwin describes as "most graceful circles" made by flying condors involves an aesthetic judgment, not derived from the life form; rather it is contingent on the taste or interest of another life form, and thus not a part of the condor's autonomous natural goodness, though flying in circles under some conditions may be. Darwin's remark about the pleasure of flying is speculation. It could be incorporated into natural history only if it could be restated in a way that allows for the special teleology of natural historical grammar (Chap. 14).

The language of natural history has a unique teleology (Chap. 10, Boxes 10.1, 10.2, 12.1, and 12.3), one that does not involve reference to a designer; it represents life forms in terms of the autonomous purposes or functions of what the bearers of the forms have and do. The naturalist discovers these only through experience. When Darwin says that the strange behavior of the guanaco occurs for some reason inexplicable to him, he indicates that his representation of the life form assumes a teleological framework that must be filled in by experience. With experience, it was later shown that the dung piles have a territorial demarcation function, among others [15]. The categorical could then be written as "guanaco males often use dung piles *for the purpose of* territorial demarcation."

Box 10.2 Convergence on the A Priori
Michael Thompson [1, 16, 17] shows that our representation of life is in part a priori: our identification and experience of living things employs a formal representation not given in any experience. Thompson shows this by describing the logical form or grammar that guides anything we think or say about living things. Experience 'fills in' the factual content of the a priori schema or template we bring to our observations of the living world. Experience and

(continued)

Box 10.2 (continued)
practice help us master application of the formal representation of life, but the representation cannot be learned or derived from experience.

Our capacity to use a priori representations like this suggests similarities with the notion that ethologists such as Konrad Lorenz called "instinct": the capacity to respond adaptively to the environment without prior experience [18, 19]. At the time Lorenz proposed this notion of instinct, the prevailing idea was that animal behavior is largely shaped by learning, and the perfecting of what is learned, through trial and error. In contrast, instinct allows animals to operate by 'trial and *success*,' [20] through a kind of a priori knowledge that gives the animal advantages in an environment which may not offer second chances. Lorenz extended the concept of innate or instinctive adaptive response to sensory perception. Some cognitive mechanisms, he argued, would be ready to function adaptively from the moment the senses generate the first perceptions. A bat does not learn to echolocate by colliding with objects.

Instinctual behaviors, innate responses to sensory stimuli, and the logic of Aristotelian categoricals are a priori, insofar as they involve a capacity to represent the world and act effectively in it, without prior experience. Do Thompson and Lorenz use the same grammar for the phrase 'a priori'? If so, the use of the language of the representation of life may be understood as a cognitively important instinct, or innate response, in Lorenz's sense [21].

References

1. Thompson, M. 2008. *Life and Action: Elementary Structures of Practice and Practical Thought*. Cambridge: Harvard University Press.
2. Wittgenstein, L. 2009. *Philosophy of Psychology—A Fragment* (trans: Anscombe, G.E.M., Hacker, P.M.S., and Schulte, J. Revised fourth edition by Hacker P.M.S., and Schulte J.). West Sussex: Wiley-Blackwell.
3. ———. 2009. *Philosophical Investigations* (trans: Anscombe, G.E.M., Hacker, P.M.S., and Schulte, J. Revised fourth edition by Hacker P.M.S., and Schulte J.). West Sussex: Wiley-Blackwell.
4. Hintikka, J., and G. Sandu. 1994. What is a quantifier? *Synthese* 98 (1): 113–129.
5. Carlson, G.N. 1980. *Reference to Kinds in English*. New York: Garland.
6. Carlson, G.N., and F.J. Pelletier. 1985. *The Generics Book*. Chicago: University of Chicago Press.
7. Darwin, C. 2001. *On the Origin of Species*. A Penn State Electronic Classics Series Publication. https://www.f.waseda.jp/sidoli/Darwin_Origin_Of_Species.pdf. See also: https://www.gutenberg.org/files/1228/1228-h/1228-h.htm. Accessed 11 Mar 2023.
8. Thomasson, A. 2019. Categories. *The Stanford Encyclopedia of Philosophy* (summer 2019 edition), ed. E.N. Zalta. https://plato.stanford.edu/archives/sum2019/entries/categories/.

9. Darwin, C. 1841. *The Zoology of the Voyage of H.M.S. Beagle. Part III: Birds*. By J. Gould (Family Vulturidae). London: Smith, Elder. https://www.biodiversitylibrary.org/item/49676#page/17/mode/1up. Accessed 11 Mar 2023.
10. ———. 1838. *The Zoology of the Voyage of H.M.S. Beagle. Part II: Mammalia*. By G. Waterhouse (Family Camelidae) London: Smith, Elder. https://www.biodiversitylibrary.org/item/124546#page/300/mode/1up. Accessed 11 Mar 2023.
11. Aristotle. 1897. *History of Animals* (trans: Cresswell, G.R.). Oxford: G. Bell. Accessed as Google Books, 11 Mar 2023.
12. Geach, P.T. 1977. *The Virtues*. Cambridge: Cambridge University Press.
13. Anscombe, G.E.M. 1958. Modern moral philosophy. *Philosophy 33* (124): 1–19.
14. Darwin, C. 1997. *The Voyage of the Beagle*. The Project Gutenberg eBook 944. https://www.gutenberg.org/ebooks/944. Accessed 11 Mar 2023.
15. Marino, A. 2018. Dung-pile use by guanacos in eastern Patagonia. *Mammalia* 82 (6): 596–599.
16. Thompson, M. 1998. The Representation of Life. In *Reasons and Virtues: Philippa Foot and Moral Theory*, ed. R. Hursthouse, G. Lawrence, and W. Quinn. Oxford: Clarendon Press.
17. ———. 2004. Apprehending human form. *Royal Institute of Philosophy Supplement* 54: 47–74.
18. Lorenz, K. 1981. *The Foundations of Ethology*. New York: Springer.
19. Tinbergen, N. 2020. *The Study of Instinct*. Cambridge: Pygmalion Press, an imprint of Plunkett Lake Press.
20. Krebs, J.R., and S. Sjolander. 1992. Konrad Zacharias Lorenz. 7 November 1903–27 February 1989. *Biographical Memoirs of Fellows of the Royal Society* 38: 210–228. https://royalsocietypublishing.org/doi/pdf/10.1098/rsbm.1992.0011. Accessed 11 Mar 2023.
21. Lorenz, K. 2009. Kant's Doctrine of the *a priori* in the Light of Contemporary Biology. In *Philosophy after Darwin: Classic and Contemporary Readings*, ed. M. Ruse, 231–247. Princeton: Princeton University Press.

11 The Form-Bearer Unity

> *[There is] a general and thoroughgoing reciprocal mutual interdependence of* vital description of the individual *and* natural historical judgment about the form or kind.
> *M. Thompson [1, p. 52; Italics in the original]*

Let us summarize the somewhat technical results of the last chapter and then illustrate some practical applications. We have seen that we 'bring' the notion of a form of life—or, more precisely, its logical schema—to our experience of living things; such experience would be impossible without it. Experiencing living things and acquiring the various notions of their specific forms is an indivisible and simultaneous process. It necessarily involves application of the generic schema of a form of life, without which it would be impossible to experience or understand the difference between being carried or pushed along by the wind like a tumbleweed, or flying through it like a bird. The specific notions that fill in the schema, e.g., the notion of flying, eating, reproducing, swimming, or making promises, cannot be generated from experiences understood somehow independently of our use of the logical structure involved. The role of this logical form is comparable to other fundamental notions that we bring to our apprehension of reality. "Quantity" plays this role for our apprehension of number, e.g., in counting or measuring extensive and intensive magnitudes. In this way, fundamental concepts—"categories" (as, going back to Aristotle, they have been traditionally called)—are a priori. Philosophers have much to say about the a priori, but for our purposes what is important is the role of language; in our view, language is the vehicle of thought, as Wittgenstein and others have held; it is guided by the grammar that we use to represent our experience of living things, among other realities. The idea is well illustrated by apprehension of the specific shape of a life encountered in some specific species, like the shape of a butterfly's life, from egg to caterpillar (larva), pupa, and adult. We learn that shape only through experience of the changes in

C. Campagna, D. Guevara, *Speaking of Forms of Life*, Fascinating Life Sciences,
https://doi.org/10.1007/978-3-031-34534-0_11

particular individual things in the world, which are understood, however crudely, to be represented by the wider context of the logic and grammar of "life forms."

Our thought and discourse about living beings, and our acquistion of the notions of their specific forms, are indivisible and interdependent activities. The individual and its form constitute a mutually and necessarily interdependent unity. As we have been emphasizing, the language that follows this logic necessarily implies certain value judgments too, those of natural goodness, a notion that we will expand in the next chapter. In describing the form of life, the language implicitly contains the standard which determines the requirements that a creature must satisfy to be a good representative of the form, i.e., a representative that is as it should be for a creature of that kind of life. The form is born by here-and-now individuals; without them there is no form and vice versa. The relation is one of mutual interdependence, like that of checkmate to chess. Without the form there is no standard for natural goodness; without the standard there is no individual creature of the kind: just as there can be no checkmate without chess or vice versa. These logical relations of natural history judgments must be understood as foundational to the practice of the conservation of life forms.

Value judgments other than judgments of natural goodness—aesthetic, spiritual, instrumental, etc., when those are not necessarily natural goodness—can overlap with those of natural goodness, but they have their own grammar that generally operates at some distance from it. We call the values involved, 'secondary values' because they do not focus on, or necessarily respect, the inseparability of form of life and individual, or the primary value of natural goodness in that relationship. For example, the notion of a form of life is not necessary to the language of the Red List, and thus no such notion is necessarily reflected in it. It does not contribute much, if anything, to our understanding of what it is to be a flying squirrel, when we know the threat category that is assigned to the species. The classification "Extinct," or any other, will be accompanied perhaps with a picture of the species, a scientific name for sure, and the population data that justify the classification, among other things. But we do not get much, if any, understanding of what a species is by reference to its classification in the list. The language of the list is quite a distance from the language of natural goodness because it operates as though form and individual bearer are not logically interconnected.

The attempt to separate form and bearer is also illustrated by talk of 'animal rights,' which applies primarily to individuals without regard to the language that describes their form, as we will see in Box 14.1 and also Chap. 24. There is no integration of form and individual bearer in the language of conservation based on these practices, as there must be in the practice of life form conservation. It is not necessarily because of how, for example, whales behave when they migrate or reproduce that determines how or whether they should be conserved. Rather, that they are sentient is the primary concern. In contrast, in the practice of life form conservation, it is impossible to conserve the form without the particular, and vice versa. And it is also impossible to make evaluative judgments about a particular without integrating its form into that evaluation.

Only when primary value is placed on the conservation of natural goodness in life, and the underlying logic of the form–particular relation is understood, can the reasons for doing conservation ground the practice of the conservation of forms of life. One might think that this notion of value is equivalent to the concept of intrinsic value. It will be seen, however, that although the term "intrinsic value" can be given a sense that is consistent with natural goodness, it is generally used in a way that is much more diffuse and that regularly diverges from that of "natural goodness" and the well-defined grammar of the language of the forms of life. In the end, though, as we hope to show, the theory of natural goodness offers us a grammar for the notion of intrinsic value that provides the notion with productive and intelligible content, in the spirit of what seems to be intended by its vague usage.

A Figure of Life

In *The Expression of the Emotions in Man and Animals*, Darwin proposes the Principle of Antithesis, which he illustrates with a picture of a dog. If the animal shows submission, or fear, the body adopts a posture antithetical to when it shows a disposition to attack or threaten. The principle applies to many other species. Without experience, one might think that the differences in posture are different *anatomies*, of different forms of life. Or, in another kind of example, given by Foot, one might mistake the sound of wind in tree foliage as originating in a living being. As Wittgenstein says: "We talk, we produce utterances, and only later get a picture of their life" [2, §224]. We have paired this with the other header quote at the start of this chapter: "How could I see that this posture was hesitant before I knew that it was a posture. . . ?" [2, §225]. Without experience, not only might angry and submissive dogs be confused as distinct anatomies but as distinct forms of life.

When the figure is understood to be a snapshot of the form of life of a dog, it can be understood to depict a posture of fear, as opposed to one depicting willingness to attack. But these distinctions in our perceptions are made possible by the way our thought and language integrates postures, or unites anatomies under a certain logic and grammar. What a particular individual *does* (the postures a dog takes here and now) is integrated with what it *is* (a dog, regardless of the posture). As the form-figure is better understood, confusion with anatomies of different forms is avoided. Confusion seems unlikely with a dog, but it is likely with other creatures, such as gorgonian corals, which superficially resemble plants, or as with the many living things that camouflage themselves. The process of understanding form and content is through the experience with the living being; we learn through experience how to 'fill out' the form or specific shape necessarily paired with a creature. Darwin arrives at his descriptions of the flying squirrel and the Principle of Antithesis through empirical observation of some squirrels and dogs, but ends up expressing a judgment about form; he arrives at the notion of "the figure of life" of each. The language builds the bridge, back and forth, between the living thing and its generic kind, for better and better representation of each.

Relevance of the United, Dual-Aspect, Character of Form and Bearer

The dominant practices of species conservation, and environmentalism more generally, are not based on the language of life forms or its logic; the practices do not advance their agendas, arguments, or policies for intervention on that basis. In conservation biology, for example, pretty much everything it wishes to express is in the language of species as populations. As we have seen and will continue to develop (Chap. 17), this concept of species has no necessary connection to the standard of life forms, and there is nothing to constrain it against treating species like 'natural resources,' among other things, having nothing to do with natural goodness. On the other hand, in practices that prioritize animal welfare, the individual is the focus of concern, and not necessarily the population: e.g., if the problem is an eagle with a broken wing, the priority is to repair the wing so the eagle can fly, not the increase in the size of eagle populations. In animal 'rescue' efforts, an injured or ill individual is assisted even when there is an abundant population. Still, the standard and reasons for intervention are not necessarily related to the form of life, but instead to issues of pain and suffering or supposed rights of the individual. Again, this is in contrast with the focus on the species as a collection, which is what is important for the IUCN Red List. The priorities behind the list reflect the view that individuals require too much investment of resources when the effect on population is low. Living things, and the forms they bear, become units of utilitarian, cost-benefit calculations, maximizing the good of the whole; or else, on the non-consequentialist views, individuals have rights that constrain concern with the whole. Neither one is focused on the standards determined by the life form-to-bearer relationship, according to which a threat to the individual is a threat to the form and vice versa.

When the form is the standard, the natural badness unnecessarily caused is reason enough to prevent the trophy hunter from killing further, for the industrial whaling to stop, etc. The interventions against such things are imperative for human natural goodness, to the extent that the human life form understands what the natural badness is and when a human being is the avoidable cause of the badness. It is a defect in the will and character of the hunter or whaler to ignore these values. Interventions in the practice of conserving life forms are to be sustained and motivated primarily by these considerations, and only secondarily from concern with population size, use value, customs, economic costs and benefits, Red List status of the species, the tastes and preferences of hunters, or any of the other typical concerns.

©Andrew Guevara

References

1. Thompson, M. 2004. Apprehending human form. *Royal Institute of Philosophy Supplement* 54: 47–74.
2. Wittgenstein, L. 2009. *Philosophy of Psychology—A Fragment* (trans: Anscombe, G.E.M., Hacker, P.M.S., and Schulte, J. Revised fourth edition by Hacker P.M.S., and Schulte J.). West Sussex: Wiley-Blackwell.

From Natural Goodness to Moral Goodness 12

In the sterility of Mars there is no natural goodness.
P. Foot [1, p. 27]

Some might wish to declare provoked extinction a 'moral evil,' [3], or to think of conservation in moral terms generally [4], but the many conservation practitioners do not agree on the particular meanings of the relevant moral terms. Their meaning, as we have noted, depends on the standards upon which moral judgments are made, and which vary according to moral views. The relevant judgments are also vulnerable, as we have seen, to being dismissed by the thought that extinction is just one of many moral wrongs [5] (although the dismissal is an avoidance or deflection of the difficulty of the reality of our annihilation of species). In short, when it comes to the provoked threat of extinction, the world of conservation presents us with quite a varied spectrum of opinion, including opposition to considering it wrong or bad. The various perspectives have led some to adopt a pluralistic sensitivity to all views, and cost-benefit analyses that would result in trade-offs among them, and from which a menu of options for intervention could be derived (Chap. 16). At best, each position would consider its reason for intervening commensurate with the others.

There is, at present, no unifying ethical perspective for conservation practitioners that could be the basis for agreement in judgments comparable to the kind of agreement humanity has managed to attain in judging murder morally wrong, or slavery or torture or theft, and other forms of categorical wrongs. We may dispute what exactly is murder or theft, in certain circumstances, but that they are moral wrongs is not disputable. Talk of "intrinsic value" is a noble attempt to take seriously more than just instrumental value of nature, and to find common ground based on the value of living things for their own sake [4, 6–7]. However, there is no, as yet, moral common ground in conservation, nor any principles or general basis for seeking it. This is particularly surprising in the context of provoked extinctions, that is in fact declared bad and wrong in a variety of language-games in conservation, including moral ones, even if they are not necessarily all commensurable with each other.

C. Campagna, D. Guevara, *Speaking of Forms of Life*, Fascinating Life Sciences,
https://doi.org/10.1007/978-3-031-34534-0_12

To hope for common ground might be futile, considering the philosophically challenging subject matter at hand: *morality*. But if agreement could be reached at least on the primary standard for evaluative judgments in conservation, then certain reasons for practicing it would be prioritized. The language of representation of life forms presents us with a standard which is decisive for judging the well-being, harm, defect, flourishing, etc., of a living thing: in a word, natural goodness or badness. The language and standard, as seen, are universal for life forms and as objective as the facts in the scientific monograph that describes any life form. Consequently, it is necessary to investigate whether a connection between natural goodness and moral goodness is possible in hopes of unifying the ethical priorities of conservation practices under the standard for life forms. We begin in this chapter to sketch such a possible connection.

The investigation of this connection takes conservation in an entirely new direction. For, in the prevailing conservation practices, the tendency has been to, instead, stretch the application of traditional morality, rather than find a standard that could unify it and the natural goodness of life forms. As we have pointed out several times, the extended application of traditional morality is what has been attempted in the animal rights or animal welfare movements. The values behind these movements are far from having unifying potential or universal application. The grammar of the language of rights, or welfare (as involving pleasure and pain, and, thus, sentience), is *in*applicable to the vast majority of life's modalities, and the application of the standards involved are typically strained applications of the standard for human natural goodness. The moral framework that is grounded in this perspective is, therefore, practically meaningless because it is both so restricted in scope of application and, at the same, stretched so thin. Alternatively, the idea of 'intrinsic value' as a basis for conservation would seem to have more universality as a basis for moral treatment of the rest of nature. However, we have already noted how its vagueness tends to disqualify it. Later, we raise other critical questions about it too (Chap. 13).

We propose that if there is to be any generally agreed-upon morality guiding the relation between humans and the great variety of other forms of life, it will be based on language with the qualities of Aristotelian categoricals. The Aristotelian categoricals express a standard for 'normative' relations, about the way things ought to be, that humans can understand by reflection on their own life form, and, with experience, in their representation of other living things. It is a language, we saw, with a dual representational-evaluative function, one that in describing the life form determines what is good or bad for its bearers: whether they are wounded, healthy, dying, sound, defective, dead, or living things in the first place. Natural goodness is a value common to all living things, derived from what they are, have, or do, according to their forms of life. The language of life forms is that which is based on a language-game for which the primary moves for the term "good" are generically the same for the whole domain of the living.

If there is a bridge between natural goodness and conservation ethics, it lies in the following thoughts: it is a necessity of our life form, i.e., of our natural goodness, to recognize and understand natural goodness and badness. This means recognizing and understanding (however roughly at times) its general presence in the living world and, thus, its instantiation in other life forms. Given this, we must ask ourselves, what does our natural goodness require of us in consideration of the natural goodness of the other life forms? Unless acknowledgment and respect for natural goodness end with human-to-human relations, then this is a question that we must confront, and it seems to be an ethical question. Coming to grips with it is part of becoming a good human being. The working general principle is this: Our interfering with the life of any representative of any form, without being able to justify it in relation to the necessities of our own form—our natural goodness—is natural human *badness*. Natural goodness and badness are not grounded in, nor interpreted, according to feelings or beliefs guided by any other standard. But what is the relationship between natural goodness and badness, and moral goodness and badness? Philippa Foot suggests an answer.

A Moral Theory Based on Natural Goodness

Foot begins *Natural Goodness* announcing it to be a book on *moral* philosophy, one that reflects on "right and wrong, and virtue and vice, the traditional subjects of moral judgment" [1, p. 1]. Moral judgment about what is good or bad, right or wrong, is usually treated as the preeminent value judgment, within uniquely human values. A main purpose of Foot's book is to argue that the standards of morality are expressed by the grammar of the natural goodness of human character and will, and thus by the same generic standard that lies in all the other forms of life, applied differentially according the specific shape of each the life. Foot thus unifies the language of moral judgment with other evaluations of the natural good and bad. The moral is a sub-category of human natural goodness. It is as much a part of the natural human goodness to be faithful to one's word as it is to walk upright on hind limbs. In doing so lies the good, and defect lies in the failure or inability to do so. If the standard of what is good for living things, in all their innumerable forms, is expressed with the language of life forms, then its dual representational-evaluative grammar should express the standard of what it means to live morally, which is part of what it is to live as a good representative of the human life form. Foot says:

> For I believe that evaluations of human will and action share a conceptual structure with evaluations of characteristics and operations of other living things, and can only be understood in these terms. I want to show moral evil as 'a kind of natural defect.' [1, p. 5]

The crucial aspect of the theory is the logical peculiarity of the language that represents living things, which Thompson develops (and that we have covered in Chap. 10). Foot's work is grounded in the results of Thompson's analysis. She writes:

> Judgements of goodness and badness can have, it seems, a special 'grammar' when the subject belongs to a living thing, whether plant, animal, or human being [1, p. 26].

Foot and Thompson are using the term "grammar" as we are: i.e., in the manner of Wittgenstein, as we have been appropriating him, and in particular with reference to the logical forms underlying the use of language. Although the natural historical grammar of the representation of living things is familiar to conservationist practitioners, they do not usually prioritize it over other types of representation and evaluation. It is ironic that the language with the potential for unifying the value of the living, across all forms, was always available, but instead subordinated to the language of use value, aesthetic value, ecological function, and so on. It is a mistake to think that these other values are the primary basis for protecting life forms, superior to that expressed by the language of life forms itself.

Foot speculates on the profound availability of the concept of natural goodness and badness:

> Some intelligent Martians who themselves did not think in terms of goodness and badness might (even if landing in the rain forest and knowing nothing of humans) realize that the plants and animals on earth could be described in propositions with a special logical form, and come themselves to talk about the newly met living things as we do [1, p. 36].

The thought is that even alien users of language could express themselves about living things by following the a priori logic of the representation of life as a fundamental category. This is because the meaning of "life" is the same at the most general level, and also therefore the meaning of "good" implicit in it. This is in contrast to the plurality of language-games used to express the values most commonly referenced in conservation—from use value to intrinsic value: they do not share a common meaning, or logical grammar, nor even follow rules that always imply consistent or commensurable values.

The philosophical defense of the theory of natural goodness as a moral theory, and its rigorous basis in the grammar of Aristotelian categoricals, was an ambitious project of Foot's till her last day, and one that Thompson has begun to develop for the concept of justice. Our ambition here is more limited, for we are not trying to reach a general philosophical audience. We wish to make the well-being of living things, and their great diversity of forms, the primary concern of conservation. For those who would question natural goodness as a basis for this, we would ask: What does it mean to be a conservationist concerned with the other forms of life, for their own sake, if not concern for natural goodness? The only further question is whether this is a moral concern, a concern with moral grounding and implications.

Foot on Natural Goodness as 'Intrinsic' and 'Autonomous'

> [N]atural goodness, as I define it, which is attributable only to living things themselves and to their parts, characteristics, and operations, is intrinsic or 'autonomous' goodness in that it depends directly on the relation of an individual to the 'life form' of its species [1, pp. 26–27].

Let us consider the essential elements of this definition, in turn.

(i) *"Natural goodness"*

Foot uses "natural" to modify the sense of "good" by reference to natural life forms on Earth. This is more specific and precise than the way it is commonly used in conservation, where "nature" can refer to ecosystems, biodiversity, the physical processes of the atmosphere and oceans, and planetary geology, among many other things. All these other natural phenomena can be associated with various standards of goodness. To borrow an example from Foot, the tributaries of a river allow the flow of the main river to be maintained. But, *if* that is in *some* sense a 'natural good,' it is logically distinct from the natural goodness of living things.

Living things are, as we saw, categorically different from rocks, storms, or sea currents. Differentiating the standards for judging what is "good," in its various senses, and what corresponds to "natural goodness" is important for the practice of conservation. The language of economics uses "useful" in the sense of "good," for both a species and an ecosystem. Aesthetic language deems "beautiful" a landscape or the courtship of a bird. Chapter 9 dealt with the issue of standards. The point here is that the norms that are installed with the language of natural goodness cannot be compromised with the norms implicit in other language-games. Hence, a conservationist of life forms does not sit comfortably at the discussion table of 'environmentalism' in its broad sense.[1] The 'environment,' like 'nature,' in its most general sense, is obviously relevant to life, but, as we have seen (Box 6.1), these are expressions which move language in directions that lose the necessary connection with the primary needs of living beings. In the justifications for a 'protected area,' for example, the language does not necessarily follow the logic of any Aristotelian categorical. Spaces can be protected from use primarily or exclusively for aesthetic reasons, for example, or for securing the opportunity for continued extractive activity. Some National Parks, early in their history, were set aside for political purposes, favoring the tastes and preferences of certain social classes, and not primarily conserved in their naturalness and relevance for creatures in them.

(ii) *Natural goodness "is attributable only to living things and their parts, characteristics, and functions"*

Individual living things are the exclusive bearers of the attributes of natural goodness, which applies, more specifically, to their parts, characteristics, and functions. The same grammar is used to represent-evaluate the bat in flight, the senses that allow it to orient itself by means of echoes, the brain cells that register stimuli, and so on. A sentence such as "the heart of the human being has four

[1]Campagna [8] discusses the justification for maintaining a distinction between species conservation and the various alternatives of environmentalism, in order to avoid the influence of environmentalist language-games generally dominated by instrumental values and human needs.

cavities, and red blood cells concentrate hemoglobin and the hemoglobin transports oxygen..." follows the form "S is, has, or does F," just as the expression "the human being uses language" does.

(iii) *Natural goodness "is intrinsic or 'autonomous' goodness"*

Foot makes little use of the term "intrinsic," or "intrinsic goodness (or intrinsic value)." We develop the relationship between intrinsic value and natural goodness in Chap. 13. In describing natural goodness as intrinsic or autonomous, Foot is concerned with the independence of the natural goodness of each life form: the exclusive dependence of a particular individual creature, on its own form (and not any other), for its primary good. This autonomy or independence is the basis of the distinction between primary and secondary value, discussed in Chap. 9. The "natural goodness" of a form as opposed to its "secondary goodness" are expressions that are grounded in distinct language-games:

> ... goodness predicated to living things when they are evaluated in a relationship to members of species other than their own, is what I should like to call secondary goodness... And we also ascribe this secondary goodness to living things, as, for instance, to specimens of plants that grow as we want them to grow, or to horses that carry us as we want to be carried, while artefacts are often named and evaluated by the need or interest that they chiefly serve [1, p. 26].

Autonomy consists in a creature's independence in the satisfaction of its life form needs, *in its natural habitat*, as opposed to dependence on other life forms for the satisfaction of those needs. The notion of autonomy, associated with that of the natural environment, serves to inform judgments about conservation actions such as assisted reproduction, translocation, and 'rewilding' (the attempt to restore altered environments to their wild condition). This concept of autonomy also allows us to put into perspective commonly used categories of threat, such as "Species Extinct in the Wild," in which "wild" refers to the natural environment of the form. When representatives of a form can no longer exist outside of the environments we have created for them, or if their existence depends on conservation actions, then their natural goodness is greatly compromised by loss of autonomy. The representation-evaluation of domesticated living beings, or those that depend on captivity, necessarily implies defects in these creatures that are hardly ever properly taken into consideration in conservation practice (see Boxes 12.3 and 14.1).

(iv) *Natural goodness "depends directly on the relation of an individual to the 'life form' of its species"*

We have discussed, in the previous chapter, the mutually interdependent relationship between individual creature and its form, and how that contains the standard for, and grounds the value expressed by, judgments of natural goodness or badness. But what is also important in the just quoted phrase is how Foot relates "life form" and "species." Particularly important is what she adds in a footnote: "I have written here of species, but it might be better to use the words 'life form,' as Michael Thompson does" [1, p. 15, footnote 14].

The difference between "life form" and "species," as the latter is commonly used, and which we have noted in preceding chapters, will be developed further in Chap. 17. The terms are used as equivalents in some contexts, but the more they are applied to the practice of life form conservation, in the context of provoked extinctions, the more we must be careful with distinctive usages. The essential thing to remember is the one we have been emphasizing throughout: in common conservation practices, especially in the context of the threat of extinction, the dominant meaning of "species" is "population." Population language is, as we have shown, driven by a grammar that deviates from the representation of life forms, even though it refers to qualities proper to sets of living things, organized on the basis of shared characteristics. It should be emphasized, however, that even population language, and thus "species," in any sense common to biology, presupposes the concept of life form, inasmuch as populations are, after all, made up of living things, and, thus, bearers of those forms. Population-oriented conservation biologists had to be proficient users of life form language before they developed the language-game of population biology.

'Aristotelian Necessity'

Foot cites Elizabeth Anscombe as a source of the concept of necessity used in the necessities or needs determined by a form of life. According to Foot, Anscombe calls "Aristotelian necessity": "that which is necessary because and in so far as good hangs on it" [1, p. 15]. Anscombe offered *promising and keeping promises* as examples of this kind of good in human life. Humans need to cooperate with each other and, without promises, they would have only force or deceit or the bonds of affection to depend on for that need—which would make human life practically impossible. As Anscombe observes:

> Now, getting one another to do things without the application of physical force is a necessity of human life, and that far beyond what could be secured by those other means[2] [9, p. 74].

The good that is involved in satisfying this necessity is natural goodness. Life form needs are all examples of this kind of necessity. Foot puts it this way:

> These 'Aristotelian necessities' depend on what the particular species of plant or animal need in its natural environment, and the ways it has of satisfying its repertoire. These things together determine what it is for a member of a particular species to be as they should be, and to do what they should do [1, p. 15].

[2]By "those other means" she refers to "exercising authority or power to hurt and help, or of commanding affection. . ."

A representative that is as it should be, has, and does what it should do, following the designs of its form, satisfies Aristotelian needs, which are expressed in the judgments of natural history. The notion of Aristotelian necessity is useful because, associated with that of Aristotelian categorical, it simplifies the expression of the foundations with which a practice of conservation of life forms operates. Following Anscombe, we call them "Aristotelian," but it is not important to our work here to explore, or verify, the relationship of these necessities to the interpretation of Aristotle's philosophy, so we speak usually simply in terms of necessities of *the form of life*, e.g., of "life form needs" or "the necessities determined by the form of life," or the like. We are concerned, as always, with the tragic irony of the threat to life forms based on poor understanding of natural goodness, the individual creature's autonomous satisfaction of the needs of its form, which are all Aristotelian necessities in Anscombe's sense. What *is* important is to recognize that it is these necessities, the ones described by natural historical propositions, that matter to the description of a life form. Foot makes this point when she notes that naturalists can identify markings, like colored feathers in their accounts of the natural history of certain birds. But unless the color can be shown to have a necessary function in the satisfaction of the needs of the bird, the natural historical propositions about it do not capture a necessity of the form, or a natural goodness (Box 12.3). These are the necessities Anscombe importantly identified and which ground Aristotelian categoricals. Only these have the right kind of necessity and teleology to support judgments of natural goodness or badness.

Conservation practices have available to them, in the theory of natural goodness, a logical grammar that can represent provoked extinctions according to the needs of each form of life, and the norms implicit in satisfying them. On the theory, the crisis of provoked extinctions is understood to be a large-scale, voluntary obliteration of natural goodness, a path that leads to the sterility of Mars, where there is no natural goodness because there is no life. The language of life forms contains the standard for categorical norms for the conservation of life forms, which do not vary according to custom or sentiment of human beings, or any other standard relative to anything but the form of life under consideration. It is a language which has the capacity for expressing, on the basis of natural goodness as the primary value, what is absolutely permissible or impermissible in our interventions, without concern for balancing costs and benefits or any concerns of what is instrumental to human beings, or in accordance with their tastes. It is, we maintain, in terms of this language that all primary reasons for conservation practice must be justified. And it is in terms of this language that there may be a moral basis for conservation, one developed in terms of a theory of virtue, as we sketch in later chapters.

Box 12.1 Categoricals

Consider the following sentences, all which suggest the form "*S is, has, or does F*:"

1. *Humans are bipeds.*
2. Humans are elegant.
3. *Humans have five fingers on each hand.*
4. Humans have little sense of direction.
5. *Humans use language.*
6. Humans wear shoes.

We have italicized the Aristotelian necessities—necessities of the life form (Chap. 12). They are here expressed as categoricals, without putting them into their teleological connections. Of those that are not categoricals, number 2 is an aesthetic evaluation, 4 is an evaluation ambiguous between two senses of "direction" (geographical or else one related to purpose in life), and 6 is based on a generalization over individuals. Unlike the categoricals, none of these express a necessary truth. Nor can they necessarily be placed into teleological relations. For some, their truth or falsity depends on opinion based on taste, or cultural outlook, unlike Aristotelian categoricals, which are necessarily and objectively true (or necessarily and objectively false, if they are false).

"*S* **is** *F*"

This component of the general notation "S is, has, does F" summarizes the teleological connections that describe what the life form S has and does. Expressions such as "humans are omnivorous," "whales are mammals," or "cycads are dioecious" summarize an indefinite number of categoricals. So, e.g., to say that "humans are omnivorous for the purpose of feeding on plants and animals" is redundant. "The whale has mammary glands that produce milk with which to feed the young" is summarized by "The whale is a mammal." And cycads are dioecious is short for individual cycads are male or female. Yet, these propositions of the form "S is F," or equivalent, serve to represent-evaluate (Chap. 8), as any other natural historical judgment.

"*S* **does** *F*"

The verb "to do" schematizes innumerable actions proper to the form of life, although not every description of what a creature *does* expresses something that properly goes into a categorical. Psychological verbs have grammars that are not those of the verb "to do" when predicated as a necessity of a life form. For example, "to strive" "to prefer," "to enjoy," or "to believe" are not necessary to the description of any life form, not even our own. It might be said that the bear prefers to spend the winter hibernating, but this derives from the

(continued)

Box 12.1 (continued)
categorical that makes no reference to preferences but instead asserts a necessary truth of the form like "The bear hibernates (i.e. enters into torpor) in the winter." "Humans prefer kept to broken promises" is necessarily true only if it is taken as equivalent to the categorical: "Humans keep promises," where the psychological verb is unnecessary. The grammar of the necessities of life forms, even human ones, is more fundamental than their psychology, and sets a priori constraints on a proper understanding of that psychology.

The schema "S does F" can be filled in by a great variety of terms, including all verbs and functions proper to the language of biological disciplines, such as "photosynthesizes," "metabolizes," "depolarizes." These will make up the propositions typical of a textbook on the life form S and its characteristic functions or activities F, giving the account of S in greater and greater detail. Likewise, many categoricals can be expressed with ordinary language, with their own level of detail:

- Plants photosynthesize in order to grow.
- Plants use the sun and air to produce materials that allow them to grow.

"S **has** F"

This schema can be filled in with a great variety of terms, rich in detail, textbook or ordinary. For example: "The flying squirrel *has* its limbs and the base of its tail united by an expanse of skin." "Humans have a heart with four chambers." Propositions that follow this structure can be placed into teleological relations.

Box 12.2 Life Cycles

> Many forms of social behavior are episodic, in extreme cases limited to narrow periods in the day, season, or life cycle. Courtship behavior and parental care, as well as the maintenance of territories and dominance hierarchies specifically linked to these behaviors, are usually seasonal.
> E. O. Wilson [10, p. 21]

In the header quote, E.O. Wilson refers to the cycles that direct reproduction, flowering, growth, and migration. Living beings go through a life cycle, which differentiates stages of development according to the age of the individual. The life cycle integrates in a unique way the seed and the adult tree, the insect in its various phases of metamorphosis, the human being from embryo to adult. Life cycles are described by Aristotelian categoricals.

(continued)

Box 12.2 (continued)
Wittgenstein observes:

> Someone says "Man hopes." How should this phenomenon of natural history be described? One might observe a child and wait until one day he manifests hope; and then one could say "Today he hoped for the first time." But surely that sounds queer! Although it would be quite natural to say "Today he said 'I hope' for the first time." And why queer? One does not say that a suckling hopes that. . . , but one does say it for a grown-up. Well, bit by bit daily life becomes such that there is a place for hope in it [11, §15].

Wittgenstein is calling attention to the fact that many of the categoricals which describe a life form apply only at specific stages of the life cycle. There is a certain relationship between the autonomy of a living being, necessary for living fully according to its form, and the stages of the life cycle. A human being in the stages of gestation is autonomous only for some life form necessities, while many of other the natural needs are satisfied by the mother. Whatever the stage of gestation, however, it is always a stage of the form of a human life.

Gestation involves asymmetrical dependence between two stages of the same form: dependence of a human in one stage (fetus) on another human at another stage (adult). The asymmetry generates possible conflict between the natural good of the gestating fetus and the gestating mother. The interruption of gestation represents an impediment to the satisfaction of primary needs of the embryo. The embryo or the fetus can also present impediments to the adult mother, including the risk of serious illness or death. Like any other ethical theory, the theory of natural goodness must address the difficult issues that arise from this conflict. As Foot says in her Postscript [1, p. 116],

> I have been asked the very pertinent question as to where all this leaves disputes about substantial moral questions. Do I really believe that I have described a method for settling them all? The proper reply is that in a way nothing is settled, but everything is left as it was. The account of vice as a natural defect merely gives a framework within which disputes are said to take place, and tries to get rid of some intruding philosophical theories and abstractions that tend to trip us up. There is nothing in the idea of natural normativity that should disturb the good work that many philosophers have recently done, for instance, on problems in medical ethics.

Box 12.3 The Necessary and the Irrelevant

> I should say that to obtain the connection between Aristotelian categoricals and evaluation another move must be made.
> P. Foot [1, p. 31]

> If one asks whether natural selection and, more broadly, all processes in evolution have a telos, one must be clear which *telos* one has in mind.
> E. Mayr [12, p. 123]

Foot points out that *not all* propositions that might be used to describe the natural history of a form of life S, according to the schema "*S is/has/does F*," identify a standard of natural goodness. She gives the example of the patch of color on the heads of certain birds, which might go into a naturalist's monograph as "S has a blue patch. . .," but which, as it turns out, plays no role in the bird's life. Darwin expresses essentially the same point in *The Origin*:

> No one will suppose that the stripes on the whelp of a lion, or the spots on the young blackbird, are of any use to these animals, or are related to the conditions to which they are exposed [13, p. 480].

There are many other features of the natural history of living things that, like color patches, spots, or stripes, neither add to nor detract from an autonomous, flourishing life. The color pattern of a zebra's skin, for example, varies with individuals, apparently without that variation affecting their ability to meet the needs of their kind of life. That is, natural goodness does not lie in one particular pattern, with the other being defective. No particular pattern is the norm. In contrast, an albino zebra is defective because "the zebra has a color pattern in which black and white stripes are interspersed" describes a necessity of the form.[3]

Foot argues that the difference is captured by the *teleological* grammar of genuine Aristotelian categoricals (see Chap. 10). When propositions schematized by *S has/does F* describe a necessity of the form, they must be linked up with each other by "in order to" clauses, clauses that describe a function or end, and that do not require any reference to a mind which designed or intended the end or function. Only experience can tell us how to fill in the schema or whether, once filled in, the resulting propositions link up with others, or how they do, if they do. But what shows that we have described a true necessity of the form and, thus, a standard for natural goodness or defect

(continued)

[3]Notice that the point is not about adaptation. Aristotelian categoricals are descriptions of what S is, does, and have, and their teleology explains function in the process of living. Adaptation is a concept that finds home in the language-game of evolution. Differences are discussed in Chap. 13.

Box 12.3 (continued)
in the individual is that the teleological connections have a unique logical grammar (Chap. 10). Teleology is then indispensable for differentiating what is necessary from what is irrelevant for flourishing and autonomous life. In short, we can take "*S is/has/does F*" to identify a standard of natural goodness if and only if experience shows that what fills it in respects the unique, teleological grammar.

Telos

"Telos" is the Greek word for "end" or "purpose." The telos of a carburetor will eventually give it away as an artifact: its 'purpose' is dependent on the ideas of a designer. Not so with the telos of living things[4]: the purpose of a natural heart does not depend on its being invented, imagined, or thought of in any way, in contrast with 'artificial hearts,' or blood pumping machines. There is no need to reference an 'intelligent designer' for the natural kind of telos. The representation of artifacts lies in the description of the design; the standards for saying what artifacts are, have, or do are determined by the design as concocted by the designer. The 'design' of a living thing is autonomous, self-contained, or 'intrinsic' to it, as is the natural goodness that satisfies the design: an autonomous and self-contained, natural value.

Darwinian adaptation

The natural life forms are of course 'designed' by evolution, through natural selection and other purposeless processes [12]. This is design without a designer, "purposiveness without a purpose," as the philosopher Immanuel Kant puts it in describing the teleology of organisms. The extension of skin that joins the limbs of the flying squirrel, and which serves it for the purpose of gliding from tree to tree, is an adaptation resulting from processes of variation and natural selection that are not the result of "conscious" or "intelligent" purposes, or the participation of a will. As Chap. 13 explains, the teleological component in Aristotelian categoricals can be thought of as a 'snapshot' of the adaptations of a life form at a point in time in the history of the species.

Domestication

Chapter 1 of Darwin's *The Origin of Species* is titled "Variation under Domestication," while Chapter 2 is called "Variation under Nature," marking the fact that domesticated forms are the result of artificial not natural selection. The teleological language that describes them must, therefore, eventually refer to the purpose of the domesticator. The plumage ornaments of some

(continued)

[4]The influential evolutionary biologist Ernst Mayr finds misleading the application of teleological language just for any end-point process. He points out that the activity of a river is not teleological just because it flows to the ocean [12, 14]. He calls "teleomatic" those processes, such as radioactive decay, that have an end-point but no goal, Teleological language fits best "genuine goal-directed processes in organisms" [12, p. 125].

Box 12.3 (continued)
domesticated pigeons, for example, do not satisfy the needs of their natural, autonomous life. The domesticated capuchin pigeon has a ruff of feathers that protrude from its head and make it difficult for the animal to see laterally. The description of this quality might be called a 'breed categorical,' but it is not an Aristotelian categorical. A good representative of this domesticated pigeon satisfies aesthetic criteria relative to human tastes and not the natural life form. In doing so, it is a defective animal with respect to the natural form described by the Aristotelian categoricals of its natural history. The language describing its domesticated form would be "the capuchin pigeon has feathers protruding from the head for the purpose of satisfying the aesthetic judgement of pigeon fanciers." Domestication, genetic engineering, and synthetic biology affect the autonomy of living things to the point of making the language of *artifacts* more appropriate to their representation than the language or life forms (Chap. 15), at the cost of ignoring an immense amount of natural goodness.

References

1. Foot, P. 2001. *Natural Goodness*. Oxford: Clarendon Press.
2. Thompson, M. 2008. *Life and Action: Elementary Structures of Practice and Practical Thought*. Cambridge: Harvard University Press.
3. Cafaro, P., and R. Primack. 2014. Species extinction is a great moral wrong. *Biological Conservation* 170: 1–2.
4. Batavia, C., and M. Nelson. 2016. Heroes or thieves? The ethical grounds for lingering concerns about new conservation. *Journal of Environmental Studies and Sciences* 7 (3). https://doi.org/10.1007/s13412-016-0399-0.
5. Marvier, M., and P. Kareiva. 2014. Extinction is a moral wrong but conservation is complicated. *Biological Conservation* 176: 281–282.
6. Rolston, H., III. 1988. *Environmental Ethics: Duties to and Values in the Natural World*. Philadelphia: Temple University Press.
7. Vucetich, J.A., J.T. Bruskotter, and M.P. Nelson. 2015. Evaluating whether nature's intrinsic value is an axiom of or anathema to conservation. *Conservation Biology* 29 (2): 321–332.
8. Campagna, C. 2013. *Bailando en Tierra de Nadie: Hacia un Nuevo Discurso del Ambientalismo*. Buenos Aires: Del Nuevo Extremo.
9. Anscombe, G.E.M. 1969. On promising and its justice, and whether it needs be respected in *foro interno*. *Critica* 3 (7/8): 61–83.
10. Wilson, E. 1976. *Sociobiology*. Cambridge: Harvard University Press.
11. Wittgenstein, L. 1980. *Remarks on the Philosophy of Psychology*. Vol. II. ed. G.E.M. Anscombe and G. H. von Wright (trans: Anscombe, G.E.M.). Chicago: The University of Chicago Press; Oxford: Basil Blackwell.
12. Mayr, E. 1988. *Towards a New Philosophy of Biology*. Cambridge: Harvard University Press.
13. Darwin, C. 1909. *On the Origin of Species*. New York: P. F. Collier. https://books.google.com.ar/books/about/The_Origin_of_Species.html?id=YY4EAAAAYAAJ&redir_esc=y. Accessed 11 Mar 2023.
14. Mayr, E. 1992. The idea of teleology. *Journal of the History of Ideas* 53 (1): 117–135.

The Value in Life

13

Biotic diversity has intrinsic value, *irrespective of its instrumental or utilitarian value. This normative postulate is the most fundamental, independent of instrumental or utilitarian value.*
M. Soulé [1, p. 731; Italics in the original]

As mentioned earlier, the main purpose of our book is practical-philosophical: our intention is to initiate a new practice of conservation, grounded on the theory of natural goodness. Some specific application of this new practice has been mentioned here and there in previous chapters; it is developed more fully in the final chapters. So far, we have advanced mostly theoretical arguments, all mainly for the thesis that conservation of living things, in all their great diversity, should rely on a particular language: the language of the representation of life, whose logical grammar has been laid out by Michael Thompson. The grammar of what he calls Aristotelian categoricals underpins the concept of a form of life—essential to any understanding of natural life—and on this same grammar Philippa Foot bases her moral theory. Foot argues that moral judgments, and judgments about the well-being (health, soundness, illness, defect) of any living beings, have one and the same general standard: namely, the satisfaction of the necessities of the forms of life. These are the necessities expressed by the Aristotelian categoricals that describe a form of life. Morality is a subset of natural goodness, a natural goodness of the will and character of human beings (or, if they exist, any other species constituted with a will, or rational agency, like our own).

In this way, we can begin to discern a bridge between the language of the representation of life and moral judgments. If we are to take seriously the judgment that provoked extinction is morally wrong or bad, the question arises whether this could be an instance of moral wrongness or badness understood as natural badness, the way Foot suggests morality should be understood In the development of these conceptual relations lies the exercise of evaluative language, particularly that which

C. Campagna, D. Guevara, *Speaking of Forms of Life*, Fascinating Life Sciences,
https://doi.org/10.1007/978-3-031-34534-0_13

has the function of recognizing and assessing the value of living things and, in a sense, life itself. The practice of the conservation of life forms evaluates provoked extinction as a natural badness in two respects: (1) in respect of its profound interference with the satisfaction of the necessities of the forms of life (threatening to eradicate the very possibility of natural goodness altogether) and (2) in respect of its doing so through intentional and avoidable acts of human will. This is where natural goodness and morality seem to connect. For it would seem to be a natural defect of our will and character to act without regard for natural goodness, which is what knowingly, routinely, and avoidably eradicating life forms evidently is.

We have arrived at these conclusions through the language of natural history, the language which describes the specific forms of life, and which implicitly contains the standard on which to base primary value judgments. The dual-function grammar of this language 'represents-evaluates' life for us: it reveals, in the factual, natural historical descriptions of it, the natural goodness intrinsic to life. This is foundational for a practice of conserving life forms: living things satisfying the needs of their form and our appreciation of the value that lies therein.

As Soulé indicates in the header quote, the idea that the diverse forms of life have value intrinsically—i.e., in themselves, and not just as a useful means to our ends—seems foundational for a practice of life form conservation. But, while Soulé's view of intrinsic value would include some instances of natural goodness, it has a different basis from that of natural goodness:

> Species have value in themselves, a value neither conferred nor revocable, but springing from a species' long evolutionary heritage and potential or even from the mere fact of its existence [1, p. 731].

Soulé finds in the evolutionary history of life the source of its value. A few lines before, he says: "Evolution is good," and he speaks of "the mere fact of its [a life form's] existence" as being good. None of this could be derived from natural goodness. Natural goodness is not a goodness that has to do with existence, as though the mere existence of a life form were good. The good of natural goodness is always relational: form to bearer.

Thus, the primary value for the practice we propose lies in creatures autonomously meeting the needs of their form. But a creature autonomously meeting the needs of its form is life itself, so natural goodness is in this way a value of—or, perhaps more precisely, *in* life itself. This bears, as we will soon discuss, some resemblance to the idea of intrinsic value, and we propose that the theory of natural goodness could provide the clarity and grounding that the idea otherwise lacks, and that could avoid certain unwanted implications. What makes for the important difference in our interpretation of intrinsic value as natural goodness is the logical grammar that underlies the latter and the language-game that grounds and motivates redress of provoked extinctions.

In the theory of natural goodness, the value intrinsic to life lies in individual creatures autonomously satisfying the necessities of the forms of their lives. This is the theoretical foundation for a concept of intrinsic value in life form conservation, one that is intimately related to the human necessity to represent life as a fundamental category. These are relational qualities intrinsic to life itself, and, in that sense, an intrinsic goodness in life. To appreciate this good, to recognize it, and to ground the actions of conservation on it is central to a conservation of life forms. The 'natural value' in life that emerges from the other creatures living according to standards of their natural good is reason enough to redress our threats to them, when those threats cannot themselves be justified by the necessities of our own life form, and this is especially so for provoked extinctions. In the language of life forms, not only are natural needs articulated across life forms, but also what is permissible, or justified, in the relationship between human beings and living beings of other forms. The permissible is found in the integration of natural human needs and the natural needs of the other forms: i.e., in the integration of human natural goodness and that of other life forms. The question of permissibility or justification only arises for *our* life form (so far as we know) because our form has certain dispositions of will and character as requirements.

In short, the reason conservation practice must be concerned with all that we have been discussing about evaluative language—the standards behind the many value judgments expressed by that language, the language of representation-evaluation of life forms, primary and secondary values, and so on—is to clear the way for a conservationist practice grounded in the good that derives from life form necessities and their satisfaction. In the practice of the conservation of life forms, what is permissible in the human relationship with non-human living things should be derived from the standard for natural goodness. Conservation practices that prioritize values other than natural goodness decouple their actions, and reason for being, from the needs of life forms. This is not acceptable for life form conservation. No standard of good overrides natural goodness. If conservation action is pursued for the good of all living things, then the language of life forms ought to guide its judgments. The practice of the preservation of life forms does not depend on, or lead to, other 'higher' values. Provoked extinctions are fostered by the subordination of primary values to secondary values.

Then, if we understand "intrinsic value" according to the language of life forms and, thus, the primary value of natural goodness, we can agree with the basic sentiment in Soulé's quote in the header. The fundamental postulate is that of the value intrinsic to life itself, in all its great diversity of forms. This value supersedes all others and cannot be traded in a utilitarian calculation of costs and benefits with the others.

Secondary Value in Life

Following Foot (see Chap. 12), we use "natural" or "primary" goodness to refer to a creature's autonomous satisfaction of the requirements of a life form, as opposed to whatever it has or does that might be good for some other reason: e.g., that it benefits creatures of another life form. However, the concepts of primary and secondary value apply broadly in many other contexts. For example, the beautiful could be considered 'primary' value for a traditional aesthetic theory, while the comfortable (an instrumental value) would be secondary to aesthetic value. When judging whether an artifact serves its purpose well, the instrumental parameter is primary and the aesthetic appeal secondary. For classical utilitarians, the primary good is pleasure, the bad being pain or displeasure. The Kantian categorical imperative has rational will as the primary value. For consequentialists, the primary good consists in the best consequences of our actions, best outcomes overall. God has primary value in traditional religion. Sustainable development is primary value for the economic instrumentalism that dominates certain conservation practices, but it is obvious that satisfying human sustainable practices cannot be a natural necessity of any form. *All* these modalities of evaluation must be considered *secondary* value for anyone who is concerned with the needs of creatures as determined by their forms of life.

We are focused in this essay on the question of the conservation of life forms: what are they, how should they be conserved, and on what grounds, for what reasons? The claim that natural goodness is *primary*, in the sense we intend, implies that no value, or reason for action, sets an equal or higher standard to the one implicit in natural goodness, i.e., the satisfaction of the needs of a life form. This is why distinguishing between secondary and primary value, as we do, is of such fundamental importance. In a sense, natural goodness is even more fundamental than moral value, since moral value is a subset of natural goodness; morality is a species-specific instance of the primary value.

On this view, we can make human benefit, at the expense of another creature, the overriding concern only when doing so is justified by the *needs* of the human form of life. These are not just any 'needs' but are typified by having to kill to eat, for example. In other words, our own benefit is primary when it is the expression of primary goodness, natural goodness. However, there is no comparable life form need to kill to the extreme of *extinguishing* another species, nor to the point of threatening extinction. And it would be hard to imagine even special circumstances in which this 'necessity' might arise. At a minimum, provoked extinction is obviously *not* justified by any need in the context of the vast majority of provoked threats and extinctions that we know of. As noted above, these are typically caused by representations of the lives of non-human forms that subordinate their natural goodness to secondary values.

The most commonly discussed secondary values in this essay have to do with the benefit to human welfare, or interests, from creatures of other life forms. Evolutionary theory also marks a related difference between primary and secondary qualities of species, as Darwin himself puts it [2, pp. 84, 183–184]:

> What natural selection cannot do, is to modify the structure of one species, without giving it any advantage, for the good of another species.
>
> Natural selection will never produce in a being anything injurious to itself, for natural selection acts solely by and for the good of each.
>
> If it could be proved that any part of the structure of any one species had been formed for the exclusive good of another species, it would annihilate my theory, for such could not have been produced through natural selection.

From the process of natural selection, we expect only that which is 'good' for the individual, that which differentially 'favors' its survival and reproduction. A 'snapshot' of this process of modification of a creature of any given species is therefore a picture of natural goodness in the form of adaptations at a particular time. A snapshot of a bird in flight, for example, or of a sturdy oak, with deep roots, standing strong in the wind.

In contrast, artificial selection, proper to domestication, subordinates primary goodness to the secondary. As we saw in Box 12.3, domestication can eliminate from the natural form of life qualities necessary for autonomously meeting needs of the form. Domestication by artificial selection is a process that makes the individual's 'well-being' or 'proper functioning' dependent on another life form (on the humans who engineered the domesticated species in the lab, or on the farm, or through cultivating a long history of dependence and instrumental service to humans: as with horses and dogs[1]).

Intrinsic Value

We have offered a friendly way of interpreting "intrinsic value" according to the language-game of natural goodness, and a way of agreeing with Soulé's sentiment in the header quote. Talk of "intrinsic value" is also found in many important places beyond Soulé.[2] Soulé himself argues that Leopold's Land Ethic reflects the "intrinsic value of nature" [1, p. 728]. We also find the term in the preamble of the Convention on Biological Diversity (CBD):

> Conscious of the intrinsic value of biological diversity and of the ecological, genetic, social, economic, scientific, educational, cultural, recreational and aesthetic values of biological diversity and its components... [3]

"Intrinsic value" is mentioned here in apparent differentiation from eight other values, all secondary to life forms. What is so special about intrinsic value that it should be highlighted in an enumeration of alternatives? Some background may help us understand what motivates the language-game in the first place.

When life is monetized, or treated instrumentally more generally, as a strategy for protecting it, there is an incentive to find potential for human use in every life form.

---

[1] Domestication may be understood as a human necessity, in, for example, the manipulation of plants that sustain the origins of agriculture. But the process may also serve 'capricious' purposes, as the case of the capuchin pigeon, but also of many dog breeds, that override basic considerations for the goodness of living things, exposing them to debilitating defects. Welfare practices are justifiably focused on the badness of the food industry's exploitation of domesticated animals, but there are other examples of badness sustained by secondary valuation.

[2] "Intrinsic value" has family resemblances with phrases such as "inherent value" or "inherent worth" (see the use of the phrase by Soulé [1, p. 731]). In practice, these and other phrases are used almost interchangeably, together with 'existence value,' for example. They do differ in some aspects, but it is beyond our purpose to analyze these differences.

But there are countless species that lack any identifiable instrumental value for humans (let alone the many that annoy us and seem worse than useless). These "useless creatures," as environmental writer Richard Conniff pointedly argues [4], have another kind of value, just as important, if not more important, for our use-obsessed society. For example, as Conniff notes: "[s]pider web silk doesn't intrigue because somebody can turn it into bandages, but because of the astonishing things spiders can do with it." The phrase recognizes the language of intrinsic value, in a kind of aesthetic language in which arguments of usefulness or uselessness are not central. However, like aesthetic value, and related values (sacredness, purity, wildness, etc.), there is considerable variability in the 'eye of the beholder,' where the strong inclination to human self-interest dominates, expressed by sentiments like this: "The only reason we should conserve biodiversity is for ourselves, to create a stable future for human beings" [5].

So we can appreciate the need for an alternative to use value, and even aesthetic value, for those who wish to value nature for its own sake, and especially to prevent provoked extinctions for their 'intrinsic' wrongness As we have noted before, it is sadly ironic that conservationists have not looked for the alternative in the autonomous satisfaction of the needs of the life forms themselves. Could there be a more fundamental reason for conservation of species than concern for those needs? It is precisely because of the things that spiders do in making webs that we should preserve them and their webs; whether they are of use *to us*, or strike everyone as magnificent, or not, is secondary. And yet, as Conniff notes, prevailing conservationist practices are always trying to justify themselves by appeal to instrumental reasons; they have more confidence in arguments having to do with the possibility of turning webs into some object for human use than they do in the magnificence of the webs themselves. The Brundtland Report does not mince words when it observes that the language of science and conservation lacks political clout, compared to that which expresses leading economic and resource concern (Chap. 6). The appeal to intrinsic value is meant to counter this, but as most commonly employed, it is an appeal to quite a disparate spectrum of values: including aesthetic qualities thought to be intrinsic (or semi-spiritual qualities like purity or sacredness). Or else, the very existence of creatures, or diversity of their forms, is sometimes said to be intrinsically valuable. This leads to a plurality of standards and, thus, of values deemed to be intrinsic, but with no unifying principle. "Intrinsic value" is a term that one finds played in many different language-games, some quite distant and disconnected from "natural goodness."

Indeed, the concept of intrinsic value has a very extensive application, far beyond that of the language of conservation biology [6–9]. Intrinsic value is central to the papal Encyclical Letter *Laudato si* [10], a religious-spiritual, environmentalist manifesto, but it is also central to a variety of other language-games. It is not exclusively applied to species and living things; the Amazon ecosystem, the Mona Lisa, God, beauty, truth, and happiness are repositories of intrinsic value according to various perspectives [6–9]. This greatly increases the challenges that pluralism presents to conservation practice [11, 12]. Finally, it is even debated whether intrinsic value is an objective value [13], as suggested by Soulé, or subjective, requiring a human evaluator to confer it [14]. The only clear common denominator in conservation seems to lie in understanding what intrinsic value is not: i.e., in the idea that living things are not just instrumentally valuable, they do not exhaust their value in the context of use. This does not say what it is about the intrinsic that deserves to be valued as a priority. Soulé refers to 'evolutionary history and potential,' but these notions do not apply in the same sense, for example, to the ecosystems for which the same value is proposed. Ecosystems do not evolve in the sense species do; an ecosystem is not a living thing, except in the sense that living things are in it. In short, "intrinsic value" is a term in wide and varied usage, played according to various language-games, all raising their own controversies and issues that we cannot expect an identifiable, consistent, unifying grammar, to resolve, in contrast with the unity of the grammar of the representation of life forms, as described by their natural history. Partly as a consequence of the controversies surrounding it, the notion of intrinsic value is often rejected on the grounds that there are more practical reasons for doing conservation than any that depend on our trying to express an abstract, rather vague, value for addressing the threats that provoke extinction. As we will try to show, the language of natural goodness grounds a conception of intrinsic value that is precise, primary, and unavoidable in any representation or evaluation of living things.

The Relationship Between Intrinsic Value and the Language of Life Forms

Ultimately, we would argue that the expression "intrinsic value" is not necessary to the language of the representation of life forms; nevertheless, we acknowledge that life form language may capture something of what Soulé and others are intuitively trying to express. We have seen that Foot (who rarely uses the term) speaks of natural goodness as intrinsic (see citation Chap. 12), in the sense of being a value found in the relation that constitutes life form and bearer, more specifically, in the independent or autonomous value of a creature's satisfying the requirements of that relationship. Then, the idea is that nothing outside of this constitutive relation counts as natural goodness. This seems reasonably close to what is sought for in the concept of the intrinsic value of life, or of the diversity of its forms.

Our only concern would be to reject the idea—common to accounts of intrinsic value—that the *existence* of living things has intrinsic value. There is a sense in which perhaps one might want to say that their existence does have intrinsic value. After all, isn't it better that there be living things rather than not; isn't a world in which there are bears and flowers, etc., rather than not, better? No doubt. But this "better" is not *natural goodness*, whatever else it may be. It is not the same thought as that contained in the judgment that it is good that a bear is hibernating, or that a plant is photosynthesizing, when these are taken as Aristotelian categoricals. For it to be an instance of natural goodness for a bear or plant to exist, we would have to be able to relate their existence to the necessity of some life form (theirs or some other), by showing how their existence is something that they have or do to fulfill the requirements of that form. For natural goodness is just a creature's satisfying the needs of its form of life, as described by the Aristotelian categoricals of the form. For existence itself to be natural goodness (and thus intrinsically good in that sense), we would have to show how the proposition "bears exist," or "plants exist," functions as an Aristotelian categorical, connecting with other categoricals that describe some life form, with the right logical grammar. This includes being able to connect such propositions with the right teleological grammar: "bears exist *so that*..." or "plants exist *in order to*...," where the connecting, purposive terms have the logic of conjunction. And it is far from clear what life form these would be describing, or whether "the in order to" clauses in such a description would have the right teleological grammar, as opposed to the grammar of intelligent design or conscious purpose, as in the theological view that bears exist in order to express the goodness of God, or the like.

In short, the notion of natural goodness is consistent with a notion of intrinsic value, without the limitations of the latter notion more generally. The language of natural goodness, as we have emphasized, is objective and exclusive to living things and their forms; its grammar, when carefully observed, does not allow for confusion with other goods, nor can it be reduced to expressing purely instrumental values. The standard of goodness lies only and precisely in a creature's autonomous satisfaction of the needs of its form of life; this standard is common to all living things, regardless of whether or not they resemble human beings, have rights, are sentient, are aesthetically pleasing, etc. The background support for this has been developed in the previous chapters, so we now can integrate all the conceptual development into these generalizations, and expand them in the remainder of the book. In the theory of natural goodness, the value of living things arises from their form of existence, but this simply means that what exists is something that is/has/does what instantiates natural goodness. That is why life form conservation does not need the notion of intrinsic value to supplement or clarify the concept of natural goodness. On the contrary, practices that are grounded in the "intrinsic value of nature" can find in the notion of natural goodness a unique and clear logical grammar to guide their use of the term.

Theories of Value Similar to Natural Goodness

There are a few theories closely associated with the language of life forms, and that even anticipate it in some respects [12, 15–17]. Paul Taylor's [17] "biocentric view" of living things as "teleological centers of life" is an important example. The biocentric view is one that understands each living thing as in a constant "teleological quest" to maintain existence and well-being. On the theory, this has the normative implication that what is good and proper for living things is that they are striving to *be* in a way not generically different from the way we strive as human beings. It may be that 'lower' forms of life, like plants, do not have emotions and other sentient qualities, but they nonetheless 'strive' toward the same general end: living and being well. It is on the basis of this that Taylor thinks all living things have "inherent worth" and ought to be given moral consideration. We achieve the biocentric view by realizing that we are just one teleological center of life, among many others, and respecting the fact that all living things have a "point of view" according to which they strive toward the good in the same way we do, and have an "interest" in attaining it; they strive toward their objective end, as much as we do toward our ends, whether they are conscious of the interest or not.

The difference between this (and similar theories; [12, 15–17]) and the theory of natural goodness is that Aristotelian categoricals do not require dependence on terms like "striving" or "points of view," or "interests," to be extended in this way, across all forms; the natural historical description of many (most) forms of life is possible (and can only be understood) without drawing from the grammar of such terms—terms which have their original home in language-games that apply to very few species. The logical grammar of Aristotelian categoricals only takes predicates from the Aristotelian necessities that describe the forms of life. Among the most general terms of life form language are "growth," "defense," and "reproduction." These general terms are found in categoricals, across forms; we cannot understand oak trees or bees without those terms. But if we try to understand life on the basis of 'effort' or 'interest,' or 'points of view,' as general terms applying to all creatures, we invite unnecessary confusion. The understanding of the natural history of the oak or the bee or any other creature is complete without appeal to these language-games, as is an understanding of their natural goodness.

Conclusions

The language of life forms that all of us begin to master at a fairly early age constantly affirms what is naturally good. Every experience with living things has the potential to add to and improve our understanding of life and its natural goodness; the innumerable ways of learning how to fill out the expression "*S is, has, or does F*" enhance our encounter with the richness of life and provide the opportunity to deepen our appreciation of its natural goodness. From the propositions of natural history there emerges a value that the user of this language understands and recognizes, at least implicitly. The use of the language of representation-evaluation can make the value more and more explicit, enriching our understanding and appreciation. From experience with living things, articulated in terms of this language, we learn how we should live, if we wish to respect the other life forms (and why we should), on the same generic standard that teaches us how and why we should respect our own. Cultural norms, personal interests and desires, feelings, or perspectives are sustained by other language-games, incapable of generating the notion of value based on the primary necessities of the forms of life. We emphasize the fact that it is ironic that these other language-games have displaced that of the forms of life, in the value judgments that guide our reasons for doing conservation. And it is tragic that practitioners, as well as the finest institutions of conservation, rely on these alternative language-games as levers for preventing extinctions, more than the language that makes it possible for us to comprehend life in the first place.

This is due in part to the common but mistaken tendency of humans to represent and evaluate by standards relevant only to their own form of life. These are often standards that are not even relevant to primary necessities, and, consequently, they motivate conservation practices according to secondary values. Indeed, the environmentalist currents that are considered most effective are those that focus on what is useful to humans. The usual alternative, non-instrumentalist perspectives, whether of the intrinsic value of all life, or of its pristine-ness, or sacredness etc., are distrusted as weak and idealistic, by the prevailing attitude. We propose that this prevailing attitude has the force and air of rationality it enjoys only because the alternative language-games concede too much to it and borrow too much from it. Again, this is sadly ironic, given the naturalistic and evaluative force and clarity of the grammar of the goodness in life intrinsic to the relation of life form and bearer, and that every naturalist and many conservationists are experts in.

The common, human-centered practices based on standards alien to primary goodness then need to be replaced by the general standard of natural goodness. That is the highest priority of the conservation of life forms. The alternative is to continue with self-defeating practices that attempt to redress the threat of provoked extinction, with values that may be the source of that very threat, like those prioritized by the Convention of Biological Diversity.

References

1. Soulé, M. 1985. What is conservation biology? *BioScience* 35 (11): 727–734.
2. Darwin, C. 2001. *On the Origin of Species*. A Penn State Electronic Classics Series Publication. https://www.f.waseda.jp/sidoli/Darwin_Origin_Of_Species.pdf. See also: https://www.gutenberg.org/files/1228/1228-h/1228-h.htm. Accessed 11 Mar 2023.
3. Convention on Biological Diversity (Preamble). 1992. https://www.cbd.int/convention/articles/?a=cbd-00. Accessed 11 Mar 2023.
4. Conniff, R. 2014. Useless creatures. *The New York Times*, September 13. https://opinionator.blogs.nytimes.com/2014/09/13/useless-creatures/. Accessed 11 Mar 2023.
5. Pyron, A. 2017. We don't need to save endangered species. Extinction is part of evolution. *The Washington Post*, November 22. https://www.washingtonpost.com/outlook/we-dont-need-to-save-endangered-species-extinction-is-part-of-evolution/2017/11/21/57fc5658-cdb4-11e7-a1a3-0d1e45a6de3d_story.html. Accessed 11 Mar 2023.
6. Vucetich, J.A., J.T. Bruskotter, and M.P. Nelson. 2015. Evaluating whether nature's intrinsic value is an axiom of or anathema to conservation. *Conservation Biology* 29 (2): 321–332.
7. Batavia, C., and M.P. Nelson. 2017. For goodness sake! What is intrinsic value and why should we care? *Biological Conservation* 209: 366–376.
8. Bekessy, S.A., M.C. Runge, A.M. Kusmanoff, D.A. Keith, and B.A. Wintle. 2018. Ask not what nature can do for you: A critique of ecosystem services as a communication strategy. *Biological Conservation* 224: 71–74.
9. Monbiot, G. 2018. Price less. The "natural capital" agenda is morally wrong, intellectually vacuous, and most of all counter-productive. *The Guardian*, May 18. http://www.monbiot.com/2018/05/18/price-less/. Accessed 11 Mar 2023.

10. Pope Francis. 2015. *Laudato si'*. Encyclical Letter. https://www.vatican.va/content/francesco/en/encyclicals/documents/papa-francesco_20150524_enciclica-laudato-si.html. Accessed 11 Mar 2023.
11. Justus, J., M. Colyvan, H. Regan, and L. Maguire. 2009. Buying into conservation: intrinsic versus instrumental value. *Trends in Ecology and Evolution* 24 (4): 187–191.
12. Sandler, R. 2012. *The Ethics of Species: An Introduction*. Cambridge: Cambridge University Press.
13. Rolston, H., III. 1994. Values in Nature and the Nature of Value. In *Philosophy and Natural Environment*, Royal Institute of Philosophy Supplement: 36, ed. R. Attfield and A. Belsey, 13–30. Cambridge: Cambridge University Press. https://mountainscholar.org/bitstream/handle/10217/37191/value-nature-updated.pdf. Accessed 11 Mar 2023.
14. Callicott, J.B. 1992. Rolston on intrinsic value: A deconstruction. *Environmental Ethics* 14 (2): 129–143.
15. Crane, J.K., and R. Sandler. 2011. Species Concepts and Natural Goodness. In *Carving Nature at its Joints: Natural Kinds in Metaphysics and Science*, ed. J.K. Campbell, M. O'Rourke, and M.H. Slater, 289–312. Cambridge: MIT Press. http://www.jstor.org/stable/j.ctt5hhj54.17.
16. Rolston, H., III. 1988. *Environmental Ethics: Duties to and Values in the Natural World*. Philadelphia: Temple University Press.
17. Taylor, P. 2011. *Respect for Nature: A Theory of Environmental Ethics*. Princeton: Princeton University Press.

14 The Value of Consciousness

I became aware of the nature of my previous life gradually,
not only through dreams but through scraps of memory,
through hints, through odd moments of recognition.
M. Atwood [1, p. 73]

We have been occupied with the way in which evaluative language frames and guides the practice of conservation. The practice of conservation of life forms that we are promoting prioritizes the value of a creature's satisfying the categoricals that describe the form of life it bears. This is in contrast to the prevailing practices of conservation, where priority is given to the instrumental value of the life form, and other types of secondary goodness: like beauty, spirituality, cultural values, among other examples. Alternatively, some practices are based on intrinsic value and we have identified the relation intrinsic to life—namely, that of life form to bearer—as the proper standard for the idea of intrinsic value. Finally, some practices find their motivation and grounds in protecting the forms of life for the sake of particular qualities generally reminiscent of what human beings value in themselves: e.g., sentience or consciousness. The value derived from possessing consciousness, especially self-consciousness, is the subject of this chapter. Conscious experience is important to conservation practices grounded in notions such as 'sentient species,' 'animal welfare,' and 'animal rights.' It is also important to the everyday human relationship with the domesticated species we love.

What Is It Like to Be a Bat?

The philosopher Thomas Nagel [2] posed this question as the title of a famous essay on conscious experience. He intended for the question to prompt us to reflect on the nature of conscious experience, not the lives of bats. The question has, according to him, no comprehensible answer for us because to know it requires the conscious

C. Campagna, D. Guevara, *Speaking of Forms of Life*, Fascinating Life Sciences,
https://doi.org/10.1007/978-3-031-34534-0_14

experiences (not just the imagination) of *being* another form of life. The conscious experience of a bat depends on perceiving the environment through sonar, for example. For a human being, it depends in part on hearing sounds of a frequency to which our form is sensitive, and which others—a dog's hearing, for example—surpass. Nagel's argument is that without having been a bat, we could not know (or understand or imagine with any reliability) what it is like to fully perceive through sonar (or without having been a dog, we could not know what it is like to hear sounds outside our auditory range, but in theirs). For us to know what it is like to be a bat would imply having conscious experiences of which we are incapable. It would require having been both human and chiropteran, and maintaining awareness of having lived as both.

Nevertheless, in *My Life as a Bat*, Margret Atwood [1] tells the story of a reincarnation, in which, with the conscious attributes of our form, one *could* know (not just imagine) one life form as another. Atwood's fictional human would truly understand the experience of bats, and so, we might imagine, have conscious awareness of experiences of many of the categoricals that bats satisfy. A character in the novel *The Lives of Animals*, by John M. Coetzee [3], suggests that if a human being can think of himself as a corpse, he should be able to imagine what it is like to be a bat, and for it to be mistreated by us.

Nagel's apparently simple thought experiment challenges the intelligibility of such stories. It is among the most widely discussed in recent philosophy and raises many questions across many fronts. But the main issue arises without our having to try to imagine what it is like to be other life forms. For example, to know what it is to experience color in the human way requires a particular sensory experience that a person born color-blind cannot have. In line with Nagel, some have argued, on the basis of such examples, [1] that the quality of conscious experience is accessible only subjectively, "from the inside;" it is not accessible "from the outside," objectively. The best available science would then be limited in explaining, or representing, the conscious experience of the individual as an objective empirical phenomenon. But then, what is consciousness all about? And if it cannot be pinned down empirically, then a question for conservation practice would be: How does one use the notion of

[1] In "What Mary Didn't Know," F. Jackson [4] presents a thought experiment: Mary, color-blind from birth, one day perceives colors. Like Nagel, Jackson concludes that all the knowledge of the physics of light, or of ocular physiology that Mary may have mastered, could not have helped her to know what it is like to see blue, red, and so on, i.e., what she has come to know only through perception. (A sign of the difficulties around this topic is that Jackson later changed his mind about this conclusion [5], but his article is still considered a persuasive classic, by many others). The illusive notion of conscious experience that Nagel and others have identified is now commonly referred to as "the subjective quality of experience," and it seems to escape any explanation in terms of the physics of light and science of vision, etc., and thus seems to be a mysterious obstacle to a scientific explanation of the world. It is worth mentioning that Michael Thompson, without having to appeal to the mysteries of consciousness, and purely on the basis of the logical grammar of language, shows that it is impossible for a physicalist language to represent the relation between life form and bearer (see Box 12.1, because it cannot express the necessary, wider, teleological context given by Aristotelian categoricals).

consciousness, which involves not only other human individuals but representatives of other forms, to evaluate good and bad with respect to other forms of life? And how does one understand how their conscious experiences make a difference to whether we have related to them beneficially or harmfully?

The Grammar of Consciousness

Until the grammar of "what it is to have conscious experiences" is clarified, how are standards of "suffering" or "well-being" to be decided? How is sentience to serve as a standard for what is good for other life forms?

Until we understand how "possessing consciousness" relates to the primary good for the forms of life that possess it, we cannot fit the term into the system of Aristotelian categoricals that describe the form. When "consciousness" is integrated into what a form "is, has, or does"—when it is translated into that language (if it can be)—it will be possible to see how attributions of it can be connected together with other categoricals by "in order to" clauses (or synonyms: "for the purpose of...," "to be able to...," etc.), which express the teleological grammar of the language of the forms of life. The difficulty that Nagel poses about accessing consciousness "from the outside" complicates the construction of that grammar. Consider, for example: "there is something it is like for *s* to perceive color so that *s* can..." However we might fill this in: how could it be used as a standard with which to evaluate the natural goodness of *s*, for any form of life *S* that *s* might bear (including, even, the human one)? How would one judge a good *s*, in this respect, if it is not possible to know what the subjective quality of its consciousness is, or if it has any? Or how would we know whether the "in order to" had the right teleology?

The problem Nagel confronts us with is not necessarily solved by the language of natural goodness. As long as access to the conscious experience of others is at issue—especially if they belong to non-human life forms—the place of "consciousness" in the language-game of natural goodness is unclear. Perhaps, though, we can work with ordinary expressions that seem to attribute consciousness, or something reasonably thought to involve it. We can say, for example, "the bat perceives and responds to the echoes of the sounds it emits in order to orient itself in space," or "the human being perceives itself and its environment through the senses of sight, hearing, proprioception, etc., in order to orient itself in space, move from one place to another, protect itself from danger, relate to other human beings, and so on." Or also: "human beings describe what they see, hear, etc. to other human beings in order to make decisions, share experiences..." This is the language of life forms and, so far as we (the authors) are concerned, it is in fact a representation of consciousness in the life form described.

However, our categoricals say nothing about what it is *like* to see, hear, etc. as a bat or any other creature—i.e., nothing explicitly about the subjective quality of experience. If Nagel and others are correct to maintain that this quality is essential to the existence and nature of consciousness, then our life form language will not have captured it. It will have captured, at best, certain functions which might not involve consciousness in the sense that includes its subjective quality. In work on 'artificial intelligence' there is debate about whether a machine could be programmed with sophisticated enough perceptual and cognitive functions to see, hear, think intelligently, etc. We might wonder whether this implies that there is also something *it is like* for it to see, hear, etc. Is the subjective quality necessary? How could a machine have subjective experiences? These are issues around the so-called "hard problem of consciousness": [6] going back to debates initiated by Nagel's famous thought experiment. Is there any such thing as the subjective quality of experience? Is it essential to consciousness? Is it really irreducibly subjective? Could machines have it? Which life forms have it? Does not having it disqualify them as sentient creatures? etc. We must leave these questions to the philosophers, and thus any comprehensive account of the role of consciousness in the natural goodness of a life.

Implications for Conservation Practice

Those who believe that the capacity for consciousness in other living things is a necessary basis for their being worthy of our ethical concern are not justified by the standards of natural goodness, unless they can demonstrate the necessity by integrating the representation of that capacity into the system of Aristotelian categoricals that describe the relevant life form. This requires understanding the objective nature of consciousness in the living things that possess it; for what a creature is, has, or does according to its form is an objectively observable matter, and the only basis for evaluating it as defective or sound, etc. Although the term "consciousness" is used in a variety of language-games—including ordinary, scientific, and philosophical language—this does not mean that it is easily convertible to the language of natural goodness. Consequently, if the argument is that ethical consideration for the other forms of life depends crucially on their conscious capacities, it remains to be seen how Nagel's question is resolved by those who make those arguments.

As we read Coetzee's novel, *The Lives of Animals*, he does not wish to make the problem of the mistreatment of animals depend on the resolution of philosophical controversies about their conscious experience. Nor do we. We have raised them only to orient our theory to them. On our theory, what is wrong or bad about our treatment of the other animals cannot depend on being able to imagine what it is like for them to experience that treatment, if that is something irreducibly subjective. So far as the theory of natural goodness is concerned, what matters is that we can say things like "The awake and healthy human being does not hallucinate" or "the healthy conscious state of the human being does not involve hallucinations." These are medical judgments, derived from Aristotelian categoricals about our sensory experiences, which allow us to evaluate hallucinations as defects or natural badness for our life form.

In the practice of conservation, the harms that humans cause to so many animals, particularly those exploited for industrial food production, matters a great deal. But the assessment of harm *according to standards of natural goodness* cannot be expressed in terms of what it is like to *feel* caged, deformed, or diminished in survival capacities. That which guides the assessment of harm according to the natural goodness of the affected forms depends not on the subjective quality of the experience of harm, but rather on the harm itself, as determined by the standards of the form. It is the standards represented in the categoricals of those forms of life that objectively ground what a harm is. And this generality applies to all forms of life equally. It applies, for example, to the genetic engineering of plants, a context in which arguments about "consciousness" presumably do not apply and yet, on the standard of natural goodness, harm is surely caused to the extent that it engineers the loss of the capacity for natural goodness, because it forces a loss of autonomy inasmuch as the plants are made to serve the interests of the engineers, another life form. It also applies in perhaps less dramatic circumstances, as with the practice of bonsai, the 'art' of inducing highly defective representatives of forms of life justified by secondary aesthetic standards. A practice of conservation focussed on sentience will overlook this harm, an illustration of its quite narrow application.

Projection of Human Standards

The focus on consciousness, or sentience, makes it easy to project the natural human good onto non-human forms and to act accordingly. This projection is common in certain conservation practices. But also, as mentioned before, and discussed in more detail in Chap. 26, a related criticism of the theory of natural goodness is that it could not possibly be good to be subjected to the aggression and violence some forms of life involve as a necessity of the form. The violent behavior of elephant seals has been cited, as we discuss in more detail later (Chap. 26).

For the time being, suffice it to say that the criticism is questionable if it assumes that pleasure and pain are, respectively, intrinsically good and bad for any form of life, including the human. This is the assumption of Peter Singer [7] and others in the animal welfare movement, where sentience is particularly important, given that only sentient beings experience pleasure and pain. On our theory, pleasure and pain must, like everything else, be evaluated by the role they play in the natural history of the life form. When Darwin described nature as "red in tooth and claw" he was paraphrasing the poet Alfred Tennyson; he was not expressing himself in the language of a naturalist. As a naturalist, Darwin would probably have represented the lion with phrases like "the lion has a mane that protects his neck from the attack of rivals." Darwin had the hypothesis that lion manes worked like a shield:

> The males of carnivorous animals are already well armed; though to them and to others, special means of defence may be given through means of sexual selection, as the mane to the lion, the shoulder-pad to the boar, and the hooked jaw to the male salmon; for the shield may be as important for victory, as the sword or spear [8, p. 102]. [2]

[2]This proposition turned out to be erroneous. The mane apparently serves for displaying reproductive quality to females [9]. Darwin did attribute the evolution of the mane to sexual selection, but in

Blood, wounds, and death are part of the risks of living, in any form of life. The lion that dies in a fight dies from defects: breakdowns in its natural functions. but that he lent himself to the fight cannot be counted as a defect. "The lion fights with other lions to assert his reproductive dominance over a group of lionesses" is a factual truth expressed as a proposition of natural history, indicating what must be satisfied as a requirement of the form, without reference to foreign standards, projected onto the lion, from another life form; the good lion fights, and sometimes dies, for reasons proper to lions, just as the good human being does, for her own reasons.

For the practice of preserving life forms, debates about consciousness, sentience, pain, and pleasure can be allowed to progress, according to the many language-games that already deal with these notions. This is another example of how our theory leaves things as they were, as we will discuss in Chap. 25. In the meantime, other standards are available for judging the wrongness of the routine harm we do to representatives of other life forms. The conservation of life forms does not depend on any quality of consciousness, but on something more encompassing and fundamental, which logically precedes such qualities, or any other supposed alternative standards and supporting language-games. Pleasure or pain, or any other sensory quality, has to be related to the whole system of categoricals that describe the form of life, and by which we judge whether a creature is being harmed or not, and how the harm figures into its natural goodness or badness. If the concept of a life form that we have been expounding is not firmly adhered to, one risks projecting the human natural good onto other living beings and thereby removing, devaluing, or subordinating objective and fundamental primary standards that allow us to arrive at a more robust and relevant evaluation of harm. The relevance of being alert to the grammar of standards is further developed in Chaps. 23 and 24.

The subjective quality of consciousness is doubly problematic as a standard for evaluation: it is problematic even in case of evaluating human goodness, insofar as philosophers and cognitive scientists still hotly debate the very existence and nature of that quality in us, and it is problematic for the common assumption that pain and pleasure are intrinsically bad and good, respectively. When it comes to the domain of the living, we must look instead to the standard in the form of life, without which we cannot tell whether we have a living thing or not in the first place. We cannot begin first with the fact that a life form is conscious, or feels pleasure or pain, any more than we can begin with the 'fact' that it is 'violent' or 'annoying' or 'repulsive' or whatnot. We begin by bringing all such facts or judgments to the standards of the form.

---

the context of male–male competition, while more recent data suggest it works better if considered as a display for female choice.

Talking to the Animals

Wittgenstein famously says "If a lion could talk, we wouldn't be able to understand it" [10, §327]. The claim is controversial and debated, like Nagel's about the bat. It will be no surprise to the reader of our essay that we side with Wittgenstein, on grounds that the only way to learn a language is through a shared form of life. This is the conclusion that interests us the most. There are attributes that find a home in one life form and not in others, as some concepts have a place in one language-game and not others. In his book, *King Solomon's Ring* [11], Konrad Lorenz maintains that a careful observation of animals can do for the ethologist what a magic ring did for the fictional King Solomon: namely, offer the ability to talk to and understand them. Lorenz did not suggest that the exchange would be based on the human language, but that it is possible to have a 'deep' understanding of animal behavior via the careful observation and familiarity with living creatures. If we want to interpret the concept of the book beyond that, we must consider all the various language-games that might be involved in doing so, and the risk of quite a bit of bewitchment too. Perhaps, a more accurate interpretation of the title of Lorenz's book would be that it would take magic to engage in a meaningful exchange of thoughts with other forms of life. Or, finally, we may just be entertaining issues about which we must remain silent.

What is plainly possible, however, is to assess whether the death of Cecil the lion—at the hands of trophy hunter Walter Palmer, in Zimbabwe, in 2015—is an act that is justified by the necessities of the human form of life. Is Walter Palmer—who killed the animal with bow and arrow, for the rush and adventure of doing so—a good representative of the human form of life or a flawed one, in this respect? The question does not revolve around Cecil's pain or his awareness of death, about which perhaps we cannot judge with any certainty. Nor does it matter whether Cecil could tell us about it, in words or roars. On the conservation of life forms the question is answered by the judgment that Palmer killed Cecil and thus gravely harmed him, without any need to do so based on the requirements of the human form of life. We have, then, an answer based on the natural goodness or badness of our form, which is in turn judged by the same generic standard, implicit in the grammar that describes the natural goodness or badness of Cecil's form of life.

©Andrew Guevara

Box 14.1: The Case of Happy
"Happy" is the name of an Asian elephant that the Bronx Zoo, in New York, received into its collection decades ago, when the animal was already 6 years old [12]. She was removed from the wild after hunters killed her mother. She then began a journey that ended in one of the world's top zoos. More recently, however, it was thought best for her that she be moved again, now from her artificial environment, to a 'sanctuary:' another, but 'better,' artificial environment. The rationale for this is that elephants belong to that handful of species

(continued)

Box 14.1 (continued)
that should be shown particular consideration on the basis of their human-like attributes: intelligence, sentience, capacity to suffer. The 'good' for these species, it is argued, involves treating them morally, which zoos cannot do. The relocation project involved a court action that ended with a contrary ruling by the New York Supreme Court. The court rejected a writ of *habeas corpus* filed on Happy's behalf. To accommodate it would have meant granting her the rights of a 'person' unlawfully deprived of her liberty.

What is most relevant to us here is that judgments about Happy's 'good' were subordinated to a moral language-game, a grammar whose home is found in descriptions of the human good, rather than the Aristotelian categoricals describing Happy's needs as an elephant. The case of Happy serves to illustrate the consequences of confusing the one language-game with the other. Her supposed human-like qualities determined what seemed to be permissible treatment toward her. The debate over Happy pushed the limits of the legal tool of *habeas corpus*, intended to protect liberty as a primary human value. It forced evaluative language to extend across the autonomy of each life form and the natural goodness pertinent to it. This strained the credulity of the courts, as it should have.

This application of moral standards to another life form is just one of many examples of the confusion caused by the dominance of homocentrism in the representation of the primary good of non-human forms. The tendency to project human qualities and standards onto Happy ignores all the serious defects she suffers on the standard given by the necessities of her own form. Perhaps some would argue that it is therefore better for her to be in a more spacious sanctuary than the confines of a zoo, given the needs of elephants. That sentiment possibly underlay the translocation project. But it, nevertheless, fails to come to terms with the significance of all the natural badness some humans have brought upon her. Happy no longer has the capacity to be what she needs to be, and a sanctuary poses its own risks, physical and psychological.

Her defective and compromised state all begins with the human act of hunting her mother, an act without reparation, affecting her forever and which continues to affect elephants in places where they still live in natural conditions. Granting her 'liberty' under the law would have changed none of this. The whole incident reflects misguided intentions to better the life of an animal suffering grave defects according to her life form needs, without grounding the sense of 'better' in the grammar of the natural goodness that meets those needs. And it might be too late for anything 'better' in that sense. A change of environment only adds further natural badness to her inability to satisfy primary needs, due to the limitations imposed by her ontogenetic

(continued)

Box 14.1 (continued)
development. And of course it does nothing to mitigate the poaching from which elephants still suffer.

"...[P]hilosophical problems arise when language goes on holiday," Wittgenstein says [10, § 38]. Talk of Happy's 'rights' and 'personhood' is an example of language on holiday, in the sense that it does not represent any particular need of the animal according to *its* form. Attention to the Aristotelian categoricals involved will show the logical confusion that. In the end, the tragedy is that it is really no longer possible to know what is good for an animal whose existence is partly a human creation, in imperfect amalgamation with nature.

References

1. Atwood, M. 1992. My Life as a Bat. *Good Bones*. Toronto: Coach House Press.
2. Nagel, T. 1974. What is it like to be a bat? *The Philosophical Review* 83 (4): 435–450.
3. Coetzee, J.M. 2016. *The Lives of Animals*. Princeton: Princeton University Press.
4. Jackson, F. 1986. What Mary didn't know. *Journal of Philosophy* 83: 291–295.
5. ———. 2003. Mind and illusion. *Royal Institute of Philosophy Supplement* 53: 251–271.
6. Chalmers, D. 1987. *The Conscious Mind: In Search of a Fundamental Theory*. Oxford: Oxford University Press.
7. Singer, P. 1973. Animal Liberation. In *Animal Rights*, ed. R. Garner. London: Palgrave Macmillan. https://doi.org/10.1007/978-1-349-25176-6_1.
8. Darwin, C. 1909. *On the Origin of Species*. New York: P. F. Collier. https://books.google.com.ar/books/about/The_Origin_of_Species.html?id=YY4EAAAAYAAJ&redir_esc=y. Accessed 2 Jan 2023.
9. West, P.M. 2005. The lion's mane. *American Scientist* 93: 226–235.
10. Wittgenstein, L. 2009. *Philosophical Investigations* (trans: Anscombe, G.E.M., Hacker, P.M.S., and Schulte, J. Revised fourth edition by Hacker P.M.S., and Schulte J.). West Sussex: Wiley-Blackwell.
11. Lorenz, K. 2002. *King Solomon's Ring*. London: Routledge.
12. Wright, L. 2022. The elephant in the courtroom. *The New Yorker*, March 7. Accessed 11 Mar 2022.

Life Forms, Artifacts, and in Between 15

The representation and evaluation of organisms artificially selected, genetically modified, or assembled by molecular engineering—all examples of prioritizing secondary value in living things—presents challenges for the language of life forms. The modification of living beings by reproductive or molecular manipulation generates organisms with the qualities of artifacts.[1] For example, there are techniques that insert genes from one life form into another (*genetically modified organisms*), or that synthesize genes in the laboratory, implanting them into an organism that in turn follows the instructions encoded by those implanted genes (*synthetic organisms*). All create organisms whose main reason for existing is to satisfy the function for which they are created. At the pinnacle of genetic engineering are artificial organisms: assembled from synthesized parts.

The Language of Artifacts

(i) *Artificial organisms*, such as a bacterium that has been genetically manipulated to produce proteins that are alien to its form, (ii) *artificial organs*, such as an artificial heart, and (iii) *inert artifacts*, such as stone arrowheads, are represented by means of a language whose grammar is similar to that of life form language, the language that represents, for example, the flying squirrel.

Consider:

> The carburetor mixes the right ratio of air and fuel in order for other parts of the engine to ignite and combust it, etc.

The proposition has a grammar that resembles this other:

> Flying squirrels glide through the air in order to travel from tree to tree . . .

[1] Our discussion synthesizes the perspectives in [1–3].

C. Campagna, D. Guevara, *Speaking of Forms of Life*, Fascinating Life Sciences,
https://doi.org/10.1007/978-3-031-34534-0_15

Both propositions are teleological and the logic is that of conjunction; the teleological grammar of the one is like that of the other, the "in order to" clause can be replaced by "and," like so:

> The carburetor mixes the right ratio of air and fuel, and the air and fuel are combusted in the engine, etc.

Nevertheless, there is a fundamental difference between the two types of teleological proposition; it lies in the difference of conditions or states of affairs under which they can be true or false. These are what standard logic calls *truth conditions*. When S = a life form (e.g., the flying squirrel), propositions of the form "*S does F in order to . . .*" are true or false *independently* of whether anyone has ever observed or had any idea of flying squirrels. For example, propositions about the flying squirrel gliding from tree to tree, etc. are true whether or not anyone has ever encountered one or formed any idea of one. This is not so when S = an artifact. An artifact that no one has ever had any idea of is not anything that has ever actually existed, nor therefore anything about which we can say anything true or false. It is at best a possible artifact. In contrast, there have been, are, and will be, many life forms, and truths about them, that (presumably) no mind has ever known, or will ever know, or imagine.

In logic, this difference in truth conditions would be marked formally by differences in what is called "the universe (or domain) of discourse," i.e., the set of objects that propositions can refer to. The universe of discourse for artifacts is restricted by the condition that the object was at least once thought up, however crudely, by some mind. The universe of discourse for an artifact, e.g., a carburetor, is restricted to objects about which some mind has formed an idea of how it is supposed to work. This condition constrains what it makes sense to think of as true or false, when forming propositions about artifacts. It constrains what can be combined as subject and predicate in a meaningful way, i.e., a way that can meaningfully be assessed as true or false. An artifact that no one has ever thought of is at best a possible artifact, about which nothing is actually, or ever has been, true or false. But, again, there are countless earthly creatures no one has ever dreamed of that actually function as they do.

This difference in domains marks a formal difference in kinds of propositions. In the cases at hand—i.e., the difference between propositions about artifacts and life forms—the formal difference reflects the logical uniqueness of the category of a life form.

As Thompson puts it:

> It is because we are finally going to have to be faced with a special form of judgment, a distinctive modality of subject and predicate in thought and speech, that I venture to say that vital categories are logical categories [4, p. 48].

Therefore, the standard implicit in the grammar for representing the forms of natural life is inappropriate for the evaluation of inert devices such as an artificial heart or a carburetor. It is very important to understand that the inappropriateness is *logical or categorical*, based on an *a priori* distinction in grammar. There are many empirically identifiable differences between an artificial heart and a natural heart, for example. We soon discover that the artificial heart must be plugged into an electric current, or requires some kind of battery to work. But the more fundamental point is that the artificial heart was concocted to work that way and would not exist unless someone had thought it up, and the standard for whether it can be judged to be a good 'heart,' or not, depends necessarily on whether it works as intended. An artificial heart and a natural one serve to circulate the blood, but to understand the first, an instruction manual is required, while the second is understood through Aristotelian categoricals. We will be playing a fundamentally misguided and confused language-game if we treat the domain of artificial objects in terms of the language of natural life. There is a categorical difference.

Imagine that we are trying to write the monograph on an artificial organism guided by the template of life form to bearer. So much of what the monograph would report about what the 'organism' has, does, and is will be understood to serve the ends of the one who concocted it—ends of another life form. In this way, the monograph would be filled with 'categoricals' that describe 'defects' or 'natural badness,' inasmuch as they would be describing an 'organism' whose 'goodness' lay in satisfying the ends of another life form, and thus secondary to the standard in natural life, which is the autonomous satisfaction of the requirements of one's *own* form. In other words, the monograph would be filled with ersatz Aristotelian categoricals.

Then, to represent artificial organisms by means of Aristotelian categoricals would force on them a grammar that would treat as defective the very functions they were built to have. Synthetic organisms, in particular, would appear to fall on the extreme end of natural badness, inasmuch as they exist exclusively to satisfy secondary needs. For artificial organisms, the categoricals would be simplified, due, for example, to the elimination of an important part of the genome, in the effort to facilitate the process of genetic manipulation. Therefore, the grammar of the representation of inert artifacts is better suited for manipulated organisms, for which the standard of how it 'ought' to be is not found in any form of life, but instead in the *idea* of the artifact that some mind dreamed up about how it is supposed to work. If one were to insist on holding artificial organisms to the life form standard implicit in their natural qualities, one would be inclined to think of them as in agony, wounded, sick, dead, etc. As would happen if the language of geometry were applied to natural human movement.

From what has been said so far, it can be understood that the value judgments appropriate to a natural living being, as opposed to an artificial one, necessarily differ. Natural goodness, as a standard of value, applies fully to self-sufficient living beings. The primary value of natural organisms lies in the form-bearer relation. It is the value of the creature's autonomous satisfaction of its life-form needs, described by the natural history of the form, and whose origins are found in its evolutionary history. Domesticated life forms, and organisms brought into the world through molecular engineering—such as genetically modified, artificial, and synthesized organisms—share, on the contrary, in their origin, interference with their autonomous life. Does this compromised natural history deprive them of natural value?

Arguments in favor of granting synthetic and artificial organisms a kind of natural goodness are supported, perhaps, by the fact that they do not entirely lack evolutionary and natural history, and some still reproduce and generate living organisms in part exposed to natural selection [5]. Moreover, it might be argued that we are still, after all, dealing with *organisms*: living beings for which there is no standard that prevents us from attributing a goodness comparable to that of natural organisms. These arguments may not establish the ethical permissibility of, or virtue in, synthesizing such organisms, but they make questionable any immediate conclusion of ethical defect in doing so. These and many other important questions would have to be addressed in a complete development of our theory.

Humanoids

The case of artificial *human* life presents the most complex and challenging questions for evaluation. How would one evaluate, on the standard of natural goodness, a golem or a Frankenstein? To the extent that we can imagine them, they would have to be evaluated as highly defective, under human life form standards. Indeed, their very status as living beings would be at stake, starting with questions about their origins. The 'family resemblances' with humans are strong, especially in sophisticated versions like the 'gynoid' in the movie *Ex-Machina*, or the computer in the movie *Her*: i.e., machines that (as we imagine them) simulate humanity so convincingly. In the cases where these humanoids have living components, the representation might require mixed grammars, in which the grammar of life forms would be allowed to guide only *some* moves in the language-game.

Here we confront the enormous area of controversy around the possibility of 'artificial intelligence.' We can glimpse the distinctiveness of our approach to the issues, one that puts the focus on the concept of a form of life. With that as the focus, our interest will be *not* in 'intelligence' or 'consciousness' or 'self-consciousness' as qualities somehow embodied in a machine or algorithm, but in those qualities as represented in a system of Aristotelian categoricals that describe a form of life. No matter how impressively the computer 'talks' to us—'muses,' 'jokes,' 'gossips,' 'solves problems,' 'shares intimacies,' with us—we will want to know whether it is a living thing or not, and (therefore) what its life form is. The impressive simulation of human intelligence and language might seem to some readers enough to reject these questions about its life form as a mere quibble or terminological issue. But consider what it is to account for a life form—human or any other—in terms of the proper categoricals and their systematic interconnection. A monograph on a creature, or what we take to be one, might go along well for quite a while before our getting stuck and having to go back to begin arranging things in a surprisingly new direction.

In any case, the 'intelligence' or 'language use' of the robot is problematic from the start. With human beings, tone of voice and facial expression matter. Is the robot defective in this regard? And what does the defect mean? Is the robot a psychopath? Is it in a trance, a hypnotic state? What is the wider context we are supposed to use to answer these questions about its form of life, or to determine whether it has one? One must really try to think through the relevant categoricals that describe human life. On what grounds will we apply or withhold similar categoricals of the humanoid, given the radical differences in origins? On what grounds can we assume that there is even a shape to its 'life,' given its unnatural origins?[2]

Extraterrestrial Life

Much the same would have to be asked about 'extraterrestrials.' Exobiology (or Astrobiology) identifies the living on the basis of characteristic features of known life forms. In Chap. 10, we saw that Thompson analyzes standard textbook lists of such characteristics and shows that they do not succeed in picking out all and only living things. The only way to distinguish the living from the inert is through the use of the grammatical representations proper to the category of life. In some contexts, it can be difficult to tell whether we have discovered a true or genuine Aristotelian categorical, but when we have, we are certainly describing a life form, and experience helps us correct the errors we might make at first. This is not so of the usual lists used to characterize the living, including anything that could be found in an image of some 'creature,' or structure, from another planet. We would have to see what the creature is, does, and has, as understood in the wider context necessary for representing its form, or how any supposed 'signs' of intelligence—a geometrical structure, say—fit into that representation.

In the search for 'extraterrestrials' it is tempting to assume that we can work with the same general concept as *earthly life form* when looking for 'family' resemblances to earthly life. But this assumes that extraterrestrial life—'life' as we do *not* know it—falls under the fundamental category of life, in ways recognizable to us because of the similarities to Earthly life. If this assumption is wrong, then the search might lead to something recognizable as similar to life but that really would have to be understood in terms of some other category, despite the resemblances to life as we know it. This may be an insurmountable impediment, as the language may not be available to represent according to the nature of this different mode of existence. Resemblances and characteristic features only go so far, and they do not necessarily determine the grammar of their description. We must know or be able to assume certain things about causal origins before we can pick out a genuine living specimen. Life-likeness, 'intelligence,' etc., if found somewhere else in the universe, raises even harder questions than those we must ask of artificial intelligence, or 'humanoids,' where we know something of origins and principles behind the patterns. The origins of life forms on earth are explained, since Darwin, mainly by

[2]Thompson's musings about the humanoid "swampman" are relevant here, especially in showing how origins make a difference [4, p. 60].

natural selection and a few other mechanisms. 'Artificial' life is explained, at least in part, by the design or algorithm of an inventor. We could not simply assume that extraterrestrial, life-like phenomena had to have one of these two types of origins. And we can only speculate about the causal principles behind any patterns we encounter, or what kind of wider context, if any, we could place them into. Forcing a system of representation, such as that of earthly life forms, on all life-like phenomena we may encounter in the universe will generate confusion. Hence, in the face of 'living' artifacts, 'humanoids,' or extraterrestrial 'life,' we cannot simply apply the standard for understanding and evaluating natural life forms on Earth. The consequences of this conclusion for the field of Exobiology require to be explored.

The Special Case of Domesticated and Hybrid Life Forms

Domesticated living beings respond, at the same time, to natural and artificial selection. That the dog evolved from the wolf means it inherited the phylogenetic history of the natural form of life, on which domestication operated [6]. Domestication leads to loss of autonomy through the predominance of the satisfaction of secondary needs (installed by the domesticator), over primary ones. From the point of view of the original, autonomous life form, domestication is a human intervention that generates a variety of deviations from natural goodness and natural selection.

It can be different with hybrids: individuals that originate from two forms of life. Hybrid forms can be natural or artificial, and may pose problems for the practice of conservation, as hybridization may play a role in evolution but also cause extinctions [7]. If the hybrid form is natural, as in many plants and animals in their natural environments, then the writer of the monograph needs to find the new categoricals, as the new form is autonomous or else selected against. That is, the categoricals that describe the new form of life are not just some integration of those from the ancestral forms. If the hybrid form is domesticated—intentionally brought about or facilitated by humans—then the language of natural goodness will be appropriate only for a portion of what it does or has. The rest could only be evaluated in terms of secondary goodness. It is possible that the writer of the monograph will start representing these domesticated functions with the language of natural goodness, but she will soon notice problems. As she applies the standard of natural goodness for ancestral species, she will inevitably have to judge certain current functions as defective. The language will vary between that of the forms of life and of secondary values. With natural hybrids, if Aristotelian categoricals from either or both ancestral species are applied, judgments of defect will abound, since the propositions that describe necessities of the original forms are not necessarily standards of the hybrid form. In sum, the hybrid form of life is a new form of life that requires Aristotelian categoricals of its own. The mule, for example, is a crossbred life form of horse and donkey. If evaluated in relation to horses, it will be found to be a very defective horse. Likewise, if evaluated as a donkey. If it is to be treated as a new life form, of its own, we will have to find the categoricals for it and, with experience, the secondary values, and the compromise of autonomy, will be made clear.

Conclusions

The language of the forms of life helps solve challenges in the comprehension of the nature of some living and inert things that might be judged 'artificial life.' The grammar implicit in the language guides primary or secondary value judgments, where they might otherwise be ignored or confused, affecting our identification and assessment of living things, important for exobiologists, for example, and including ethical issues that can arise. Artifacts perfectly engineered to function as natural, autonomous living things may fool the naturalist in composing her monograph—as may life-like extraterrestrials with origins unlike anything we know. We are not infallible, but we do have a reliable and rigorous standard guiding us, through our ordinary language.

References

1. Preston, C. 2008. Synthetic biology: Drawing a line in Darwin's sand. *Environmental Values* 17 (1): 23–39.
2. Sandler, R. 2012. The value of artefactual organisms. *Environmental Values* 21 (1): 43–61.
3. Coyne, L. 2020. The ethics and ontology of synthetic biology: A neo-Aristotelian perspective. *Nano Ethics* 14 (1): 43–55.
4. Thompson, M. 2008. *Life and Action: Elementary Structures of Practice and Practical Thought*. Cambridge: Harvard University Press.
5. Sandler, R. 2012. *The Ethics of Species: An Introduction*. Cambridge: Cambridge University Press.
6. Morey, D.F. 1994. The early evolution of the domestic dog. *American Scientist* 82 (4): 336–347.
7. Allendorf, F.W., R.F. Leary, P. Spruell, and J.K. Wenburg. 2001. The problems with hybrids: Setting conservation guidelines. *Trends in Ecology and Evolution* 16 (11): 613–622.

Pluralism 16

Together, we propose a unified and diverse conservation ethic; one that recognizes and accepts all values of nature, from intrinsic to instrumental, and welcomes all philosophies justifying nature protection and restoration, from ethical to economic, and from aesthetic to utilitarian.
H. Tallis and J. Lubchenco [1 , p. 27]

We sometimes speak, generically, of conservation, as though it were one thing. But the multiplicity of evaluative standards in conservation indicates that there is not just one conservation practice but many that variously understand what it means to 'do conservation' [e.g., 2–6]. Conservation practices differ in the allowable moves of their various language-games. The conservation of life forms follows the grammar of natural history judgments, which represents-evaluates the form-bearer relationship and is implicitly guided by natural goodness as the primary standard of goodness. Other conservation practices, we saw (Chap. 9), differ in their standards of goodness, so they operate with a variety of value systems. Consequently, the reasons for doing conservation in each practice, e.g., for planning and intervening to fix what is judged wrong or bad, differ according to the moves allowable in the language-game, in ways that are not necessarily compatible (as discussed in Chaps. 23 and 24; see also [5]). And yet the willingness among conservation practitioners to adopt or encourage *pluralism* is widespread, involving a variety of language-games licensing a diversity of possible evaluations of the same state of affairs. As one would expect, we here approach 'pluralism' as a plurality of language-games.

The openness in conservation to a plurality of language-games of representation and evaluation of nature and life, as an 'ecumenical' strategy [7] for taking on the challenges of provoked extinction, ignores how profound differences in judgment can result from the way language-games constrain, or liberate, our use of a disparate mixture of terms and concepts. This chapter argues that, despite its good intentions, pluralism is not a viable strategy, and in fact the attempt to install it is self-defeating because the standards work against each other, when not guided by the overarching

C. Campagna, D. Guevara, *Speaking of Forms of Life*, Fascinating Life Sciences,
https://doi.org/10.1007/978-3-031-34534-0_16

primary standard for life forms. We are of course in favor of being open-minded and respectful to a variety perspectives. Our concern is when pluralism generates conceptual or logical conflicts that go unaddressed, or when certain popular perspectives dominate while others, very fundamental ones, are lost [5].

Grammatical Cacophony

Our criticism of pluralism seems out of step with the prevailing view in conservation, which tends to hold that, in the absence of unity of discourse and judgment about what is good in conservation, plurality is best because it can select from the strengths in each view. We see the view reflected in the titles of many publications [1, 7–9]: "*Reconciling Conflicting Perspectives for Biodiversity Conservation in the Anthropocene*," "*A Call for Ecumenical Conservation*," "*A Call for Inclusive Conservation*," "*Ethical Pluralism, Pragmatism, and Sustainability in Conservation Practice*." The header quote, along with the following, illustrates the language of pluralism in these documents:

> We propose a particular "practice" of conservation—one that tolerates all ideologies . . . ; we are confident that conservation will be more successful if it embraces all motivations and ceases to act as an arbiter of moral purity . . . ; conservation needs an openness of its perspectives and a practice adapted to context [7].

What kind of perspective would emerge from merging ideologies? The pluralist perspective hopes that common cause can be enriched by embracing a variety of evaluative standards. This ignores the fact that language itself can present us with real difficulties, when seeking agreement in values, since the language used to express the values variously grounds, and is grounded in, standards that may be irreconcilable, either because they conflict or are incommensurable. For example, when evaluating trophy hunting, the IUCN argues on instrumentalist grounds that the activity should be permitted, if done sustainably and with an eye to preventing corruption, ensuring compliance, etc. [10]. Indeed, as IUCN experts see it, "[a] trophy hunting programme can serve as a conservation tool" if properly managed [10, p. 7].

However, these judgments are incompatible with the individual animal's *well-being*, in pretty much any sense of the term. And the conservation community is split on this issue in ways that are irreconciliable without compromise. And for some any compromise would be unthinkable. The organization *In Defense of Animals (IDA)* condemns recreational hunting in the strongest possible terms, as "the murderous business," saying that:

> Hunting is a violent and cowardly form of outdoor entertainment that kills hundreds of millions of animals every year, many of whom are wounded and die a slow and painful death [11].

Unless the perspectives and work of organizations like IDA, *The Humane Society*, *Born Free*, or *International Fund for Animal Welfare (IFAW)* are unjustifiably ruled out as a "practice of conservation," it is hard to see an ecumenical view emerge in which recreational hunting is good in any sense. Examples like this abound. To pick another, shark finning would be judged as equally reprehensible by many conservationists. But in the context of 'sustainable' fisheries, shark finning is permissible under the rules of the so-called 'administration of nature,' which insure perpetuation of the activity on condition of 'abundant' shark populations or compliance with particular procedures, such as landing the entire animal, not just the fins. Moreover, even when conservation biology shows that sustainable development strategies are failing, the finning does not necessarily come to a halt, if it can be argued that the economic interests of fishermen are what must be sustained. With a plurality of values to choose from, 'sustainability' is a moving target in this style of conservation. These approaches are impossible to reconcile with the values that emerge from the standard implicit in the form-bearer relationship, hence with the practice of conserving life forms.

Considering what has been said about language and values, when the different language-games of conservation practice bring incompatible or incommensurable standards to the evaluation of the same state of affairs, there is a predictable cacophony of language, confusion rather than productive articulation of the points of view. How, for example, can points of view that flatly reject the concept of the 'pristine,' calling it a failed metaphor of conservationists with a 'nostalgia for wilderness' [12], be made compatible with others that idealize it and use it as inspiration for their practices? The difficulties posed by the differences in language-games are insurmountable, in part simply because they involve standards that cannot be meaningfully compared. Pluralism assumes that it is possible to select what works best from the variety of perspectives [9]. But a competitive selection process, among a Babel of grammars, can make it very difficult to identify one form of evaluation as superior to the others, which is necessary for avoiding competition between incompatible value systems once it is agreed that there are incompatibilities. And, as we have pointed out many times, a language-game can succeed in installing the value system facilitated by it, for example by investing more financial resources, without thereby making any essential contact with the forms of life which it is supposed to represent.

As we see it, contrary to certain influential views [6], many of the practices of conservation today are not just diverse, in some innocent and desirable way, but profoundly divided; they are better examples of conflicting forces than converging forces: 'eco-pragmatists' versus 'naturalists' who hold to 'intrinsic value'; 'sustainability advocates' versus devotees of the 'pristine'; 'ecological economists' versus 'deep ecologists,' to mention but a few. Some currents of thought, such as those committed to economic development—including the language-games of 'natural capital' and 'ecosystem services,' for example—are fostered by national governments, which install these language-games through control of the resources necessary for environmentalist practices, particularly by favoring the work of certain NGOs that play the game. Consequently, the language peculiar to these perspectives tends to appropriate the whole field of standards for representation and evaluation. Grammatical resistance through competing language-games is almost non-existent. Worse, some NGOs preemptively adopt the same language-game as governments or corporations, with a view to accessing funds that require this stance [13]. But while a plurality of perspectives may seem to improve conservation options, the diversity of standards of evaluative judgments exacerbates the difficulty of making all the language-games compatible. The differences, as hunting and shark finning illustrate, function as real barriers despite those who would regard and treat them as simply compatible differences. The differences often go as deep as the *logic* of representation and of correspondingly incompatible or non-comparable values.

Language and Cognition

According to pluralism, the language of intrinsic, spiritual, or religious value is as welcome as that of instrumentalism, aesthetic appreciation, or animal rights, among many others, for representing and evaluating threats to species, including provoked extinctions. In practice it means that, in all contexts related to environmental issues, we must weigh and adjudicate a formidable spectrum of competing values having to do, for example, with everything from the effects on poverty, the agricultural frontier, sustainable development, overuse, mega-mining, climate change, political will, etc. Moreover, the value judgments supported by all the various language-games involved must not only be understood by all points of view honored, but each point of view must recognize the reasons for action, motivated by those judgments, as perfectly appropriate. This expectation will be frustrated by the obstacles that grammar sets to the limits of thinking or making sense, in a given language-game, and which some conservation practitioners will be unwilling to accept as viable, unwilling to use, or unwilling to admit as an alternative. Finally, the attempt to make compatible the values that are associated with divergent language-games will encounter the difficulty of inertia; it is easier to stick with the language-games that are most familiar to us, especially when they are well funded.

Recall the passage, cited early, by the historian of science Lily Kay:

> O]nce the commitment is made to represent life in a particular way—at the material, discursive or social level—that representation takes on an entity that enables, but also constrains, the thoughts and actions of biologists. In a way, it is representation that guides imagination and reasoning [14].

The language of the representation of life is the point of departure for everything else in the work before you. To represent with language is to condition a perspective; it is to think and it is to act. There is research by cognitive scientists, psychologists, and linguists that shows the influence that language has on cognitive and perceptual processes [see 15–17 for some examples]. The results indicate that the preferential use of certain words or expressions, to the exclusion of others, frames and influences our understanding of things [17]. If a particular term expresses a concept in one language, while another language does not contain such a term, even if the concept can be explained in other words, this can lead to differences in the perception of what the concept expresses [15].

These are interesting and important results. And the empirical evidence from the cognitive sciences reinforces our view of language as *not* a passive form of representation that must adapt itself to any state of things in order to represent it properly. However, the theory of natural goodness is *not* grounded in experimental results of this kind; it is instead based on the insight that the concept of life is a fundamental logical category, whose logical grammar is indispensable to identifying anything as living in the first place. The representation of life employs an *a priori* schema or template we must apply to the empirical world in order to apprehend the reality of natural life, and to evaluate whether any living thing is doing well or badly. This is the basis for our taking natural goodness as the primary value, and the language of life forms as the unique and indispensable language-game, in the conservation of life forms.

Every language user is limited by the grammar guiding her use of language and the grammars available to her are limited by the cultural and historical situation in which she has learned the language-games of her own particular idiolect [18]. Likewise, one who has practiced conservation based on the spiritual comfort that comes from appreciating nature will encounter, in trying to understand the eco-pragmatists, a deeper obstacle than the one that lies in the glossaries they each draw from. Conservation based on the pragmatic attitude in slogans like, "whatever works," assumes that users of different language-games can objectively judge the relative effectiveness of assessments and grounds for interventions from among competing language-games. How could they otherwise agree on what it means for an assessment to 'work'? Let alone that it works better than their own!

In its proper place, trophy hunting may 'work' for some eco-pragmatists, but, we saw, it is impossible for it to be considered an acceptable conservation solution for others. In a framework that imposes (often by dint of political and economic power) values that allow animals to be treated as a means for human entertainment, the fight for the animals' welfare against the trophy hunters starts at a great disadvantage, one that can make its position seem naïve and weak. Arguments in favor of protecting animal welfare will tend not to seem as effective as those based on support for salaries for park rangers, without which poaching might destroy the value of the protected area. This assessment, however, does not take into account that the value of animal welfare arguments can hardly get traction in a scenario where such instrumental standards dominate. The instrumentalist language-game constrains thought and action in a way that we would find appalling in other contexts, very close to hand. Imagine what it would be like to play the instrumentalist language-game with human pets. The idea of weighing the costs and benefits of pet trophy hunting would have little opening advantage; it would be generally considered to be itself an appalling expression of value, and we would wonder if proponents knew what the full significance of "pet" was.

The Ideal of a Common Grammar

Ecumenical conservation practices have the ambitious aspiration of finding a language-game in which "natural capital" and "pristine" would be terms assessable according to one overall, integrated standard. What happens in practice, however, is not integration but summation. It seems at times as though the way the aspiration is thought to be fulfilled is with a list containing all imaginable values in one sentence, assuming that leaving none unnamed is sufficient for synergy to occur. The example of the CBD preamble is again useful:

> Conscious of the intrinsic value of biological diversity and of the ecological, genetic, social, economic, scientific, educational, cultural, recreational and aesthetic values of biological diversity and its components . . . [19]

This pluralistic declaration does not present us with an integrated evaluation. It is a simple menu of options that illustrates the difficulty involved in confronting the heterogeneity of values. Even if there is a will to integrate, the difficulty is logical. Integrating logically diverse language-games is like playing with pieces from different games in a new game that no one knows, or is even willing to imagine, if what is required is to set aside the language that one finds most familiar and natural. At the very least, the imagined pluralist language-game will be far removed from one that places primary value in living things autonomously satisfying the needs of their forms.

In short, we misunderstand the workings of language when we expect that better conservation practice will result from finding agreement among language-games that involve a variety of standards for making value judgments, not necessarily compatible with each other. The problem of shark finning is not best solved by understanding it from the framework of food chains associated with the cruel treatment of animals, associated with pleasure in luxury consumption, associated with international trade, and so on. And, much less, it is likely that we will converge in judgment, when at the discussion table each one of us plays her own language-game in the best way she can, before attempting to learn to make the moves in competing language-games, which she is least prepared to do. Indeed, a pluralist attitude may have to admit that, for some perspectives, finning is not even an issue. Each language-game will evaluate according to the way it represents the relevant states of affairs. The exercise of trying to integrate a variety of language-games may at least serve to demonstrate the distance, or similarity, between them, but, inevitably, confusion in the reasons to conserve is the eventual result.

References

1. Tallis, H., and J. Lubchenco. 2014. Working together: A call for inclusive conservation. *Nature* 515: 27–28.
2. Soulé, M.E. 1985. What is conservation biology? *Bioscience* 35 (11): 727–734.
3. Kareiva, P., and M. Marvier. 2012. What is conservation science? *Bioscience* 62 (11): 962–969.

4. Doak, D.F., V.J. Bakker, B.E. Goldstein, and B. Hale. 2014. What is the future of conservation? *Trends in Ecology and Evolution* 29 (2): 77–81.
5. Vucetich, J.A., D. Burnham, E.A. Macdonald, J.T. Bruskotter, S. Marchini, A. Zimmermann, and D.W. Macdonald. 2018. Just conservation: What is it and should we pursue it? *Biological Conservation* 221: 23–33.
6. Sandbrook, C., J.A. Fisher, G. Holmes, et al. 2019. The global conservation movement is diverse but not divided. *Nature Sustainability* 2: 316–323.
7. Marvier, M. 2014. A call for ecumenical conservation. *Animal Conservation* 17 (6): 518–519.
8. Kueffer, C., and C.N. Kaiser-Bunbury. 2014. Reconciling conflicting perspectives for biodiversity conservation in the Anthropocene. *Frontiers in Ecology and the Environment* 12 (2): 131–137.
9. Robinson, J.G. 2011. Ethical pluralism, pragmatism, and sustainability in conservation practice. *Biological Conservation* 144 (3): 958–965.
10. IUCN Species Survival Commission. 2012. *Guiding Principles on Trophy Hunting as a Tool for Creating Conservation Incentives*. Gland: IUCN. https://portals.iucn.org/library/efiles/documents/Rep-2012-007.pdf. Accessed March 11, 2023.
11. In Defense of Animals. https://www.idausa.org/ Accessed March 11, 2023.
12. Marvier, M., Kareiva, P. and Lalasz, R. 2012. Conservation in the Anthropocene: Beyond solitude and fragility. *Breakthrough Journal*, Feb 1.https://thebreakthrough.org/journal/issue-2/conservation-in-the-anthropocene. Accessed March 11, 2023.
13. Mark, J. 2013. Naomi Klein: "Big green groups are more damaging than climate deniers". *The Guardian*, September 5.
14. Kay, L.E. 2000. *Who Wrote the Book of Life? A History of the Genetic Code*. Stanford: Stanford University Press.
15. Caldwell-Harris, C.L. 2019. Our language affects what we see. *Scientific American*, January 15. https://www.scientificamerican.com/article/our-language-affects-what-we-see/. Accessed March 11, 2023.
16. ———. 2019. Kill one to save five? Mais Oui! *Scientific American* 25 (5): 70–73.
17. Boroditsky, L. 2011. How language shapes thought: The languages we speak affect our perception of the world. *Scientific American* 304 (2): 63–65.
18. Costa, A., A. Foucart, S. Hayakawa, M. Aparici, J. Apesteguia, J. Heafner, and B. Keysar. 2014. Your morals depend on language. *PLoS One* 9: e94842. https://doi.org/10.1371/journal.pone.0094842.
19. Convention on Biological Diversity (Preamble). 1992. https://www.cbd.int/convention/articles/?a=cbd-00. Accessed March 11, 2023.

Species and Forms of Life

17

The process of natural selection, acting in every population, generation for generation, is indeed a mechanism that would favor the rise of ever better adapted species.
E. Mayr [1, p. 135]
[A]lmost everything we think of an individual organism involves at least implicit thought of its form.
M. Thompson [2, p. 64]

Equipped as we are now, with the concepts of grammar, language-game, Aristotelian categoricals, teleology, natural goodness, and form of life (or life form), among others, and having set out the unique logic and grammar of life form language, we are in a position to focus on certain differences that arise in the use of the terms "species" and "form of life" (already noted as important in Chaps. 4 and 7).

The last paragraph of *The Origin of Species* [3] is the most famous of the most famous book on species of natural life, and it was written by the same author who described flying squirrels in language that is a perfect example of the use of natural history propositions to describe a form of life:

> There is grandeur in this view of life, with its several powers, having been originally breathed into a few forms or into one; and that, whilst this planet has gone cycling on according to the fixed law of gravity, from so simple a beginning endless forms most beautiful and most wonderful have been, and are being, evolved.[1]

[1] This citation belongs to the first edition of the Origin [3]. The digital book is not paginated, but the paragraph is the last of Chap. 14. In later editions, Darwin changes the beginning of this paragraph to this version: "*There is grandeur in this view of life, with its several powers, having been originally breathed by the Creator into a few forms or into one…*" [4 , p. 529]

C. Campagna, D. Guevara, *Speaking of Forms of Life*, Fascinating Life Sciences,
https://doi.org/10.1007/978-3-031-34534-0_17

Darwin speaks of "forms," describes them as innumerable, admirable, and beautiful, and finds grandeur in a conception of life in which forms develop and change, and do not remain eternal as time flows on a geological scale:

> In considering the *Origin of Species*, it is quite conceivable that a naturalist, reflecting on the mutual affinities of organic beings, on their embryological relations, their geographical distribution, geological succession, and other such facts, might come to the conclusion that each species had not been independently created, but had descended, like varieties, from other species [4, p. 22].

"Species," "variety," "geographic race," and "subspecies" are terms Darwin uses for the sets of individuals that share an evolutionary origin and form of life. Darwin, like Mayr in the header quote, must think in terms of populations, since evolution by natural selection cannot be explained except in those terms.[2] However, Darwin also uses the phrase "form of life" (or, alternatively, "life form") and more or less synonymously with "species." As, for example, in saying,

> ... Natural Selection almost inevitably causes much Extinction of the less improved forms of life ... [4, p. 23]

> ... a large continental area, which has undergone many oscillations of level, will have been the most favourable for the production of many new forms of life, fitted to endure for a long time and to spread widely [4, p. 119].

The origin of species is the origin of forms of life and—as we have seen in his account of the flying squirrel, etc.—Darwin understands forms of life in terms of their natural historical descriptions. Then, Thompson, Foot, and Darwin all use "species" and "form of life" more or less interchangeably, and as understood in terms of their natural historical descriptions. However, as already cited in Chap. 12, Foot warns: "I have written here of species, but it might be better to use the words 'life form' as Michael Thompson does ..." [6, p. 15]. And about species and forms of life (or life forms), Thompson says:

> I will be using the words 'life-form' and 'species' more or less equivalently in what follows, with some ambivalence noted as the occasion arises. The latter expression is used in empirical science and might reasonably be given over to it, but it should be remembered that the English words 'form' and 'species' arise from philosophy, in particular from Latin translations of Aristotle's *eidos*. The principal difference between them, from the present point of view, is in associated ideas: in thinking of a particular species, I will imagine a manifold of individuals outside and alongside one another; in thinking of a particular

[2]Darwin himself requires us to consider species as populations to explain the gradual nature of evolutionary change in life forms (this gradualism could not be easily explained by competing notions of species; see [5]). We have been arguing that the use of population language in conservation departs from the language of life forms, but we recognize its essential and primary role in evolutionary biology. As we go on to explain, this does not change the fact that life form language is more fundamental, presupposed by any notion of a population of living things.

> life-form, I will imagine one individual, the image having the standing of, say, a picture in a field guide [7, p. 28].

For Thompson, "species" and "life form" both belong to a logical category represented by the particular language of natural history propositions. Consistent with Wittgenstein's practices, Thompson fixes his meaning not by giving a definition, but by his use of the relevant terms [7, p. 62], and a description of their grammar. "Form of life," as we have seen, refers to what can fill the place of the subject position, *S*, in the particular expressions of natural history that are the Aristotelian categoricals: "*S* (the form of life) *is/has/does F*." The grammar is such that the predicates which apply apply necessarily.

Foot's concept of natural goodness derives from these necessities of life forms (as noted in Chap. 12). She says,

> Natural goodness, as I define it, which is attributable only to living things themselves and to their parts, characteristics, and operations, is intrinsic or 'autonomous' goodness in that it depends directly on the relation of an individual to the 'life form' of its species [6, p. 26–27].

> ... Natural-history sentences, which Thompson also calls 'Aristotelian categoricals', speak of the life cycle of individuals of a given species [6, p. 29].

And also,

> ... it is the particular life form of a species of plant or animal that determines how an individual plant or animal should be ... [6, p. 23]

The passages coincide with Thompson's view that "[w]hat merely 'ought to be' in the individual we may say really 'is' in its form" [7, p. 81]. In speaking of the "form of life of a species" we take Foot to mean "species understood as a form of life," and to be signaling that "species" is not always understood this way in biology [8–10], as we have pointed out already, emphasizing the implications this has for evaluative judgments in the conservation of species. "Species," in the sense of a "form of life," is not described by the species-as-population language of biology; it is what is described by the relevant set of Aristotelian categoricals. Only the latter are essential in this context, since the value judgments about living things that we are necessarily concerned with are based on them. Our purpose here is to relate "species" and "forms of life" in a way that can help prevent misunderstandings that arise from the grammar of population biology, which has appropriated the concept of species.

Species and Forms of Life in the Context of Conservation and Environmentalism

Conservation practices cannot be the same when the goal is to conserve life forms as opposed to species as populations. We have provided and will continue to provide arguments in support of this point. For example, we have discussed how in the language of populations, an extinction is described as the complete absence of particulars, population zero. From the point of view of the language of life forms, all the essential information is ignored or lost in this representation of extinction: all that a living being of a certain kind is, has, and does. This is to lose sight of what, from the point of view of life form conservation, motivates and grounds the practice—its goals and interventions—in the first place.

It is the technical sense of "species" as populations, used in the empirical sciences, which is the main focus of conservation practices falling under "conservation biology" and "conservation science," but also in the broader conceptual framework of environmentalism. Populations represent living things according to the relationship between a set and its members, in a certain surrounding and within certain parameters, depending on the purposes of the relevant empirical investigation. As we (the authors) see it, the environmental movement tends to use this framework in its focus on threats to human beings as a threat to human 'ecology' [11]. This is at least the way the words "ecology" and "environment" are commonly used by governmental and intergovernmental agencies, such as those in the United Nations and the United Nations itself. The Brundtland Report was the work of the Brundtland Commission of the UN [*aka.*, the World Commission on Environment and Development, whose focus was human beings. Typical environmental threats include droughts and floods attributed to climate change, pandemic-causing viruses such as COVID-19, contamination of air, water, soil, food, and also lack of drinking water or sewage treatment. The language of environmentalism is highly influenced by governments, which in the last 30 years have tended to deal with species under the umbrella of the Convention on Biodiversity and the discourse of sustainable development.[3] In this grammatical environment, shark finning would be assessed as a sustainability issue and an opportunity for trading, in accordance with practices that represent species as populations: as long as sharks numbers do not drop significantly, the activity is 'good'. A species does not get classified as threatened by the Red List until there is expert evidence that population sizes, or the distribution of those populations, are affected by shark finning. Absent from such classifications are representations of living things by what they are, have, or do according to their forms. Even where the language of natural history is used, it is not what grounds the reasons for inclusion in the list (it is not because the flying squirrel glides from tree to tree, and similar features, that it receives its category in the list).

Dancing in No Man's Land [11] argued for the desirability of differentiating species-focused conservation practices from homocentric environmentalist language. This differentiation should lead, it was argued, to our observing a sharp distinction between environmentalist and species conservationist practices. The distinction is justified by the effect of the differences in language, and in the way the grammar of the human-centered environmental movement tends to neutralize alternatives, and to prioritize certain values over others, and, consequently, certain

[3] An exception would be the *Endangered Species Act* of the USA, a legal framework in which species are not subject to exclusively secondary values [12].

interventions and the evaluative judgments of right or wrong, good or bad, permissible or impermissible, which motivate and ground them. And now, on the basis of the central concepts of the work before you, we propose that practices of conservation of species as populations be sharply distinguished from those based on species as forms of life, a critical step for the practice of life form conservation. This will help ensure that when an extinction occurs, the representation of it will include all that natural history teaches us about the loss of the life form, rather than a crude oversimplification of it, based on quantitative parameters.

Conservation of Species as Populations (as Opposed to Forms of Life)

The Red List is the only objective and qualified source of information on the threat posed to tens of thousands of species in their global distribution. The IUCN Species Survival Commission, responsible for the list, represents species according to taxonomic groups. For each group, experts assess the conservation status of the species concerned. Each species assessed is classified according to ecological and population criteria of abundance and distribution. And so it is that the world builds an idea of the conservation of life and its diversity, on quantitative population parameters, like 'abundance' (Chap. 7), size, mean, percentage, rate, or trend of population. Examples of this would be: "all flying squirrels live in forests in the northern hemisphere," "one in 5000 human male newborns has hemophilia," "few giraffes are albino." These are quantitative generalizations referring to individuals in a population. As we have seen, these generalizations are quite different in logic and grammar from the life form propositions of Aristotelian categoricals (Chap. 10, Box 10.1).

As we have pointed out many times, especially in Chap. 15, population generalizations do not use the standard of life forms for identifying or evaluating individual creatures or the status of a life form. The language of populations has no *direct* or *primary* concern with natural goodness or badness. When all that matters is 'abundance,' the albino giraffe counts as an individual in its population, undifferentiated from the one with species-typical skin color. From an evolutionary perspective, this individual may be selected against, and that because of how it performs in life. But what is accounted for by evolution is the effect on the gene pool of the population, to which the albino may not have contributed. On this standard, natural goodness matters only indirectly, insofar as it contributes to the likelihood of an increase or decrease in the numbers. So, it is a *secondary* value on this way of evaluating how a species or its members are doing.

When species are understood as populations, then a 'good' conservation practice is one that keeps that population within a certain measure of the relevant parameters: e.g., fertility, age-class mortality, and other quantitative group standards, not affected by human action. A practice is 'bad' when interventions do not prevent population declines which may be caused, e.g., by high pre-reproductive mortality or low fertility. The language of conservation, at the scale of populations, thus integrates quantitative generalization and evolutionary variation. In terms of distribution, the spatial continuity of groups, or the geographic range, is also often important. This aspect of species conservation is therefore often handled by a particular practice, that of 'conservation of spaces,' which is discussed in the next chapter.

Conserving 'Species' Does Not Ensure Conserving Life Forms

The fundamental differences that exist between the many conservation practices, for which the distinction between species and forms of life is relevant, will occupy the last part of this essay. Suffice it to say here that the concept of life forms is a problem for the practice of species conservation based on population parameters. The concept poses, as we have seen, a necessary, logical connection between form and bearer: the idea of form is not conceivable without that of bearer and *vice versa*. The individual matters because it necessarily bears a form of life, not because it contributes to a quantifiable fact. These differences, in substance and grounds, lead to logically distinct and often incompatible value judgments that motivate and justify the differences in particular practices of conservation in the first place. That is why they are important to this discussion.[4]

For example, as we have discussed (Chap. 4), there are few representatives left of the vaquita, the endemic dolphin of the Gulf of California, and, due to certain fishing practices, their population decline does not cease [15]. It has been suggested, as an intervention, that all representatives still living in their natural environment be removed and kept alive in captivity [15]. One aim is to avoid immediate extinction until a better protected natural area is found. If this intervention were to be carried out, it might allow dolphins of this kind to continue to exist for a while longer than they otherwise would. This conservation achievement would result in highly defective individuals, greatly compromised in their autonomous capacity to satisfy primary natural needs. But so long as the quantitative parameters did not show a population of zero, the act of 'conservation' (removing them from the wild) might be judged desirable. It is not that the ability to live autonomously, in accordance with the form of life, does not matter to those proposing this sort of intervention, but that it cannot be primary consideration, given everything and the urgency of saving the population. 'Survival' comes first. Similarly, so-called 'animal culling' may require killing a proportion of individuals of a population to maintain abundance. It does not necessarily matter whether the individuals being 'extracted' are themselves flourishing representatives of the form.

[4]When 'conservation science' [13] was distinguished as conceptually different from 'conservation biology' [14] the species-life form distinction was implicit; population talk was common to both but the language of conservation biology was closer to the notion of life forms than that of conservation science.

The intervention in the Gulf of California was abandoned as the procedure caused the death of some captured individuals [16]. Many arguments both ethical and pragmatic were adduced on its behalf. None were explicitly or necessarily grounded in the natural goodness of the dolphins. On the standard of natural goodness, the only possible justification for our putting the dolphins into captivity would be some necessity of our own life form, as it obviously cannot be a necessity of the dolphin form to be removed by humans from its environment, since the necessities of a life form are only those that can be satisfied, autonomously, by a bearer of the form. There are no other justifiable interferences on the standard of the autonomous satisfaction of the needs of its form. On that standard, it is very hard to see what human life form need could possibly ground the proposed interventions. Could it be a moral necessity arising from the fact that human beings are responsible for the dolphin's dire straits in the first place? We will return to the question of specific moral implications like this in Chap. 19. However, we propose straightaway that our energy and efforts are better directed to understanding and addressing the original causes of our routine and avoidable interferences with natural goodness, rather than continuing to perpetuate them and the imposition of values secondary over the primary value for life forms, with our misbegotten interventions. We are aware that, for the vaquita, this may guarantee the extinction of the form. If that is the case, then at least life form grammar would represent more accurately the loss than any alternative, including, of course, population language. When no justifiable intervention is possible, i.e., when no intervention based on the standard of natural goodness is possible, we should at least come to terms with the loss and address the natural badness that led to it.

To sum up: when the practice is one of conserving species as populations, the 'primary value' is measured by population parameters; good is abundance, permanence, resilience, recovery, growth, expansion; bad is decline, extinction. When the practice is based on the needs of the life form, what is primary is the individual's autonomous satisfaction of those needs. The language will express all such needs and also, therefore, the permanent loss, in extinction, of any possibility of their being satisfied. Abundance and the rest may be important perhaps as proxies for protecting life forms and avoiding their provoked extinction, but in the language of life form conservation they are understood to be only signs or indicators of the main value and focus of conservation. Conservation interventions based on these indicators only address the signs rather than the real problems, and the interventions proceed as though species and specimen are separable for the purposes of conservation, based on separate standards. But as the language and logic of life forms show, individual creature and form are necessarily integrated and interdependent. The individual is crucial; what it is, has, and does cannot be irrelevant, or treated as a negligible effect on the parameters of a set. It is the bearer of the form without which there would be no life forms, or life at all, to circumscribe by population parameters. The final chapters of this essay contain further development of these points.

References

1. Mayr, E. 1992. The idea of teleology. *Journal of the History of Ideas* 53 (1): 117–135.
2. Thompson, M. 2004. Apprehending human form. *Royal Institute of Philosophy Supplement* 54: 47–74.
3. Darwin, C. 1859. *On the Origin of Species*. First Edition. See in: https://www.gutenberg.org/files/1228/1228-h/1228-h.htm (Release Date: March, 1998 [eBook #1228]. Accessed March 11, 2023.
4. ———. 1909. *On the Origin of Species*. New York: P. F. Collier & Son. https://books.google.com.ar/books/about/The_Origin_of_Species.html?id=YY4EAAAAYAAJ&redir_esc=y. Accessed March 11, 2023.
5. Mayr, E. 2002. *What Evolution Is*. Basic Books.
6. Foot, P. 2001. *Natural Goodness*. Oxford: Clarendon Press.
7. Thompson, M. 2008. *Life and Action: Elementary Structures of Practice and Practical Thought*. Cambridge: Harvard University Press.
8. Mayr, E. 1976. Species Concepts and Definitions. In *Topics in the Philosophy of Biology*, Boston Studies in the Philosophy of Science, vol. 27, 353–371. Dordrecht: Springer. https://doi.org/10.1007/978-94-010-1829-6_16.
9. De Queiroz, K. 2011. Branches in the lines of descent: Charles Darwin and the evolution of the species concept. *Biological Journal of the Linnean Society* 103: 19–35.
10. ———. 2005. Colloquium Ernst Mayr and the modern concept of species. *PNAS* 102 (1): 6600–6607.
11. Campagna, C. 2013. *Bailando en Tierra de Nadie: Hacia un Nuevo Discurso del Ambientalismo*. Buenos Aires: Del Nuevo Extremo.
12. Endangered Species Act of the United States of America. 1973. https://www.govinfo.gov/content/pkg/COMPS-3002/pdf/COMPS-3002.pdf
13. Kareiva, P., and M. Marvier. 2012. What is conservation science? *Bioscience* 62 (11): 962–969.
14. Soulé, M.E. 1985. What is conservation biology? *Bioscience* 35 (11): 727–734.
15. Rojas-Bracho, L., R.R. Reeves, and A. Jaramillo-Legorreta. 2006. Conservation of the vaquita *Phocoena sinus*. *Mammal Review* 36 (3): 179–216.
16. Rojas-Bracho, L., F.M.D. Gulland, C.R. Smith, B. Taylor, R.S. Wells, P.O. Thomas, B. Bauer, M.P. Heide-Jørgensen, J. Teilmann, R. Dietz, and J.D. Balle. 2019. A field effort to capture critically endangered vaquitas *Phocoena sinus* for protection from entanglement in illegal gillnets. *Endangered Species Research* 38: 11–27.

18 Conservation Without Life Forms

When I say "I'm not a biodiversity guy,"
I say "I'm an ecologist."
P. Kareiva (In: [1, p. 60])

With particular attention to the 'conservation of spaces,' this chapter discusses conservation practice as sustained by judgments that are not primarily guided by the standard of a creature meeting the necessities of its life form, the standard of natural goodness. As the language of life forms becomes more and more irrelevant to the outcomes of conservation interventions, they become less and less dependable, from the point of view of the needs of the life forms and, thus, natural goodness. Interventions resulting from these practices might happen to generate benefits for life forms, but they might just as likely be useless from the point of view of natural goodness—particularly, in preventing provoked extinctions.

The practice of the so-called 'conservation of spaces' is ultimately concerned with conserving living things but not from direct focus on species as life forms. A 'space' in the intended sense may be an object of conservation within the language-game of ecology (e.g., a protected habitat) or within a language-game of spatial relations treated as a fundamental category (e.g., a 'transboundary' area)—no population or life form language required. Likewise, a wetland, or a submarine canyon, may be an object of conservation *per se*. Living things are present in these spaces, of course, but what species are there is not a *primary* concern in justifying conservation interventions. Under the concept of 'spaces,' a relevant quality of species is their *distribution*. As the language reflects, spaces ('environments,' 'habitats,' and 'ecosystems') are the objects of conservation practices that are thus *distinct* from those of practices that seek to protect abundance and diversity, or from other environmentalist and conservationist practices, including, of course, life-form conservation. The crucial point for us is that the language-game of the conservation of spaces can proceed without even so much as mentioning

C. Campagna, D. Guevara, *Speaking of Forms of Life*, Fascinating Life Sciences,
https://doi.org/10.1007/978-3-031-34534-0_18

the Aristotelian categoricals that describe the forms of life in those spaces. This is so even though the declared purpose might be to save the living things in those spaces.

Conservation of spaces is, therefore, a specialized practice, supported by a specialized language-game. This is illustrated by the fact that the IUCN, for example, has the World Commission on Protected Areas, distinct from the Species Survival Commission. 'Protected areas' are central to the conservation of spaces, even if the underlying reason would be said to be the protection of the living things that depend on these spaces; in this language-game, a National Park is a typical example of a protected area [2]. Nevertheless, the reason we are here giving special attention to the conservation of spaces is that we see the possibility for integrating it with the conservation of life forms, for all the differences between them, and we briefly make suggestions at the end of the chapter for doing so. Just to give a hint of the idea here, there are Aristotelian categoricals that incorporate the habitat (or portions of it) in their description of the format.

The Language of the Conservation of Spaces

As noted, the language-games that underlie the practice of creating protected spaces are different from those that underlie species conservation, although it is generally assumed that both ultimately have the purpose of redressing threats of species extinction. As we have seen, the vocabulary common to this language-game of "spaces" includes terms like: "environment," "habitat," "site," "place," "territory," "area," and "ecosystem" where innumerable populations, and forms of life, are to be found. The terms will be understood according to their use, but it is not necessary for us to be precise about that here. The point is that "spaces," in the various senses of the term, do not directly focus on the language of the forms of life, and often diverge from it.

We can cite familiar examples of spaces for our purposes. A forest can be understood as a space, so too a jungle or coral reef, a coast, a desert, an island, the seabed, a plateau. Terms such as "jurisdiction," "sovereignty," "boundary," "private or public property," and "commons" are commonly applied to spaces in planning and executing interventions. The concept of the pristine applies to spaces and it is important here too, as are the concepts indicating its violation. Spaces can be planned, degraded, or restored. All these terms and concepts are available for the assembly of arguments on behalf of the value of spaces for conservation without reference to life form functions or population abundance, etc.

To the extent that the conservation of spaces does, at least indirectly, involve the protection of species-populations, the language of spaces has, from early on, made bridges to that of species-populations. Yellowstone was the first National Park in the world, and its creation was justified in this way:

> Congress' principal purpose in creating Yellowstone National Park was to preserve the geysers and hot springs of the region and to protect the herds of bison, elk, and other wildlife that inhabited the park [3].

Talk of protected areas for species is a language that conservationists, in general, understand. However, arguments in favor of creating the areas do not necessarily depend on species protection as part of their rationale. Areas are conserved, for example, to protect ecological functions or processes, trophic relationships, or physical qualities of an environment. Again, species underlie this all, though reference to them is not to be found as a focus of the language. For example, regarding the expansion of the Papahānaumokuākea Marine National Monument, Sally Jewell, the US Secretary of the Interior during Obama's presidency, declared,

> The Northwestern Hawaiian Islands are home to one of the most diverse and threatened ecosystems on the planet and a sacred place for the Native Hawaiian community ... President Obama's expansion of the Papahānaumokuākea Marine National Monument will permanently protect pristine coral reefs, deep sea marine habitats and important cultural and historic resources for the benefit of current and future generations [4].

Consequently, as already suggested, an area may be protected even without much knowledge of the diversity of species it supports—it may be protected for ecological, spiritual, and aesthetic reasons, among others. But the point is that neither (i) the language of species, understood in terms of population size, abundance, etc., nor (ii) the language of life forms, based on the distinctive grammar of "*S is, has, or does F*," when S = a form of life, is predominant or essential, when the goal is to create a protected space. Neither the language of species as populations nor of life forms is *in*compatible with the language of spaces, and both commonly accompany the language of spaces, but the latter tends to overshadow both, with that of quantitative measures unrelated to populations or life forms, such as surface area or geographic boundaries expressed by latitude and longitude.

Then, despite the implicit understanding that a protected space protects living things, there is a tendency for the language of conservation of spaces to stray from that of the conservation of species or life forms, keeping only indirect or generic connections. The Mission of the IUCN World Commission on Protected Areas reads:

> To develop and provide scientific and technical advice and policy that promotes a representative, effectively managed and equitably governed global system of marine and terrestrial protected areas, including especially areas of particular importance for biodiversity and ecosystem services [5].

An assessment like "the rainforest provides ecosystem services to humankind" is a typical example of a move in the language-game of spaces—of rainforests, in this case—alienated from the forms of life that depend on the relevant environment to meet the needs of their forms: needs which have now gone entirely unexpressed. It is true that the same or similar outcomes might result once in a while from interventions based on spaces and species or life forms; e.g., ecosystem services might dovetail with life form needs sometimes. But there is nothing to guarantee the result or even make it reliable. It is possible to create, for example, a particular area for the management of industrial fishing and, at the same time, to consider it a protected space, since the fishing activity is limited within it during periods agreed to by the industry. This kind of intervention can be completely appropriate for some pragmatic approaches to the conservation of spaces, but of course it will be insufficient, *in*appropriate, and even *un*acceptable from the perspective of other practices of conservation aiming, for example, to conserve biodiversity. The intervention is certainly a move hard to justify by the language-game of the conservation of the forms of life, where natural goodness is the primary guide for practice, rather than wildly different standards of 'sucess.'

Still, one and the same practice might be able to accommodate all three perspectives: conservation of spaces, life forms, and species-populations, although moving from one to the other requires changing the language-game. For example, the language of spaces might be integrated into Aristotelian categoricals, in this way:

> The Gulf of California dolphin lives in coastal waters in order to feed on prey found at shallow depths.

Coastal and shallow waters are marine spaces. But now, this must be contrasted with the following:

> The northern Gulf of California is a dangerous place for dolphins because of totoaba fisheries.
>
> The dolphin population inhabiting the northern Gulf of California is critically endangered.

As we have illustrated already (e.g., Chap. 4), these last two are examples in which a space is *not* necessarily or directly related to the primary needs of the form of life, even though the language allows for evaluations that lead to species protection, understood as population protection. When a species is designated as Critically Endangered, the evaluative judgment is based on the standards of population biology, where abundance is prioritized. But this does not explicitly describe or involve anything about the needs of the life form. That shark finning involves cutting fins in an unsustainable manner says nothing about sharks, their form of life. And the setting aside of a 'no-take' protected area, in order to prevent shark finning, is reasonable, but detached from the language of the forms of life. We emphasize: the language-game of the conservation of spaces may assume that its interventions foster the meeting of many life form needs, but it is a language-game whose interventions are not necessarily justified on the basis of those beneficial effects. The reason to practice conservation of spaces is not directly and necessarily grounded in the values of life forms. The northern Gulf of California became a protected area because dolphin numbers were dropping significantly, a justification based on the conservation of populations, and one that does not require justification in terms of the language of life forms.

The protection of spaces based on justifications that are not guided by the language of natural history, or the Aristotelian categoricals it expresses, allows for reliance on 'multiple-use protected areas.' This alternative of the conservation of spaces allows certain uses of the 'protected' space—such as fishing for certain species—as long as only authorized fishing gear is used, or catch limit is not exceeded, for example. Of the Papahānaumokuākea Marine National Monument, the US Department of Commerce has said:

> The Department is committed to protecting ecosystems like the Papahānaumokuākea Marine National Monument for future generations, and we are working with commercial fishermen to safeguard the continued economic vibrancy of this industry [4].

So we have the following moves in the language-game of conservation of spaces: (i) the protection of space for the sake of its diverse and threatened ecosystem, (ii) for the sake of its sacredness for the Native Hawaiian community, (iii) for the sake of protecting pristine coral reefs, (iv) deep sea marine habitats, (v) important cultural and historic resources, and, finally, (vi) in order to safeguard the continued economic vibrancy of the fishing industry. All of the above benefit current and future generations—all good intentions, no question, but no language representing the primary goodness of living things. What it is that these spaces are supposed to do for life in the ocean is not explicitly expressed. Then, so far as the language goes, only, or mainly, secondary values are applied to the living things that depend on the protected area (except perhaps humans).

Once again, we see that as the language diverges from that of life form language, interventions based entirely on secondary values for life forms are permitted. The primary needs of the potentially captured living dolphins do not have priority among the expressed reasons motivating the protected space. Life form needs are fulfilled at best indirectly, by, for example, not allowing reproductively active individuals to be fished, out of concern for stability of the population. We have no doubt, particularly in the dolphin case, that the final purpose of the conservation of the space was to rescue the living things from a predicted death. What we are doing here is not to question motives, but to provide a clear description of what is really happening through the sort of attention to language that Wittgenstein locates in an inquiry that:

> ... sheds light on our problem by clearing misunderstandings away. Misunderstandings concerning the use of words, brought about, among other things, by certain analogies between the forms of expression in different regions of our language [6: §90].

Conservation Ecology

In the quote at the head of the chapter, the ecologist Peter Kareiva, a proponent of 'conservation science' working hand in hand with corporations, proposes an alternative to Soulé's 'conservation biology' [7], by distinguishing between *biodiversity* and *ecology*. Conservation practices of the latter type are focused on communities, ecological processes, environments, ecosystems, or populations. The latter type of practice focuses on species, of course, but the language may well allow referring to them indirectly, via population parameters, the interdependency of communities, the variety of spaces where they are distributed, the energetic relations that linked them in food webs, etc. Thus, the conservation interventions of an ecologist like Kareiva are not based exclusively or primarily on protecting species for the values implicit in their natural history. How, then, is provoked extinction to be understood by this perspective?

We saw that species-based conservation practices represent and evaluate extinction through language-games centered on the notion of population. For the language of ecology, the concepts of abundance and distribution are critical. Other equally important concepts have to do with processes like those involving nutrient cycles in an ecosystem or a food chain. But in the end, our assessment, from the point of view of the practice of life form conservation, is that they all tend to leave out the essential thing. While these other practices do not entirely ignore, or disassociate their practice or interventions from, life forms—in fact, they all must implicitly and indirectly rely on life form concepts insofar as they are talking about living things—they nevertheless do not explicitly or directly prioritize life forms. They ignore or submerge the standard implicit in the relation of life form to bearer and prioritize or invent other standards. There is rarely if ever explicit use of representation in terms of Aristotelian categoricals or the standard of natural goodness implicit in them. One can, for example, focus on language that expresses the relevance of primary and secondary 'producers' and 'consumers' in the 'food web,' without even mentioning species. This is perfectly proper ecological language. But comprehensive ecological views also integrate the domination of the human being into the picture, as urged by Kareiva and others—taking seriously the possibility that standards of human goodness might be detached from the natural goodness of any other form of life: human exceptionalism. This facilitates the dominance of anthropocentric, instrumentalist values, tending to the assessment of provoked extinction in terms of economic costs and benefits. Life forms have, in this context, no relevant priority.

The anthropocentric-instrumentalist conservationism, which 'eco-pragmatists' such as Peter Kareiva, Michelle Marvier, and Robert Lalasz [8], have argued for, can be articulated without any reliance on the language of representation-evaluation of life forms [9]. It can also adopt an ideological political stance that rejects as obsolete certain notions of the conservation of spaces, such as those that prioritize the notion of the pristine:

> ... conservationists will have to jettison their idealized notions of nature, parks, and wilderness—ideas that have never been supported by good conservation science—and forge a more optimistic, human-friendly vision [8, p. 2].

On the eco-pragmatic view, the pristine, we saw (Chap. 16), is an unproductive ideal, unfriendly to human beings. For the eco-pragmatist, the function of conservation is to facilitate a rational instrumentalization of nature, rather than to pursue 'impractical' ideals of the purity of nature untouched by human beings:

> Conservation should seek to support and inform the right kind of development -- development by design, done with the importance of nature to thriving economies foremost in mind [8, p. 4].

Replacing conservation based on the intrinsic value of nature—or of practices critical of the instrumentalization of life forms, ecosystems, etc.—with this 'new environmentalism' requires, for the eco-pragmatists, a new frame of mind according to which the existence of human beings contributes to the value of nature and poses no real threat to a strong and resilient nature:

> The notion that nature without people is more valuable than nature with people and the portrayal of nature as fragile or feminine reflect not timeless truths, but mental schema that change to fit the time [8, p. 4].

The wording is a good example of terms that run amok, but with assertions like this, eco-pragmatists mean to dominate what counts as rational discourse in conservation practices [10]: what is rational is the pro-human perspective, in which all other life forms are subordinated to the realities of the Anthropocene; the new mindset is acceptance of human dominance and not necessarily human natural goodness. The language of life forms is irrelevant or, at least, unnecessary as a standard, and perhaps even detrimental, by not speaking in a way that seems relevant to the powers-that-be. The pragmatic standard of conservation prioritizes human 'needs,' but with no necessary connection to the Aristotelian categoricals for our life form. Keeping the economy growing is not an obvious necessity of our form, though it may be desirable from the point of view of eco-pragmatism. And, if by prioritizing human 'need' we finish off endangered species, we can rest assured in the knowledge that nature is 'resilient.'

In sum, eco-pragmatism is a reactionary rejection of a traditional school of conservation biology, for which the intrinsic value of 'pristine' and 'wild' nature and non-human intervention in ecosystems are primary values. Eco-pragmatic ideology subordinates the biocentric view to an anthropocentric and developmentalist one:

> The conservation we will get by embracing development and advancing human well-being will almost certainly not be the conservation that was imagined in its early days. But it will be more effective and far more broadly supported, in boardrooms and political chambers, as well as at kitchen tables [8, p. 5].

These ideas are not that different from those expressed in the Brundtland Report, some decades before eco-pragmatism tried to establish itself as the most realistic and reasonable current in conservation. It is, therefore, not really offering us a new environmentalism, but the one which has always had the powers-that-be behind it. One that sees provoked extinction as best addressed by proper management of nature, the human-friendly administration of the unprecedented annihilation of non-human life forms. For these reasons, among others, the eco-pragmatic view of conservation has been vigorously criticized and opposed by many central players in conservation biology and ethics [11–16].

In short, the various conservation practices we have been considering are grounded in essentially different language-games. Eco-pragmatism and life form conservation *are in*compatible. But the others are not entirely incompatible with life form conservation: they only differ in the relative importance they give to natural goodness. But if we keep priorities straight, they can be integrated. For example, the language of spaces acquires particular force when it is incorporated into Aristotelian categoricals:

> Woodpeckers hammer **a hole in the tree** that they use as a nest to incubate their eggs.
> The wolf defends a **territory large enough** to secure prey to hunt.
> Whales migrate across **wide expanses of ocean** to reach feeding and breeding grounds.

Every monograph of a life form will have some Aristotelian categoricals that describe the characteristic habitat or environment of the life form. In this way, life form language must make reference to spaces, such as those we have highlighted in bold. Some examples, such as the hole in the tree, may be too specific to be included among the list of spaces to be conserved by a practice of conservation of spaces, but, on the other hand, even that level of detail could be relevant to a management plan aimed at promoting the satisfaction of necessities. Such spaces would be included as part of a larger habitat, like certain forests. But we have grasslands, wetlands, and the deep sea, among many other spaces, regularly incorporated into the language of Aristotelian necessities of the species dependent on them. Box 18.1 schematizes possible integration across all three: conservation of spaces, populations, and life forms.

Our concluding message is the same. The ironic dominance of conservation practices other than one based on the language of life forms is possible only through inattention to the grammar necessarily guiding any representation of the living world. The Red List, for example, is an invaluable contribution to population-based conservation language, but it operates at quite a distance from life form needs, and little apparent awareness of the consequences for conservation interventions. With no sense of the logical or grammatical implications, the language of quantity—'abundance,' 'rarity,' etc.—dominates and displaces the language that expresses what living things are, have, or do, according to their form. One consequence is the remarkably rapid emergence of the eco-pragmatism of Kareiva et al., facilitated by the focus on secondary values and presented as a new, more reasonable, approach to our relationship to nature, when in fact it is a retrograde reversal of the all-to-recent struggle to address the crisis of provoked extinction driven by human-centeredness.

Box 18.1 Three Conservationisms

	Species conservationisms	Space conservationisms	Conservationism of life forms
Important terms and phrases from the language-game	The language of the Red List The language of animal welfare The language of conservation biology, population dynamics,	Area, habitat, boundary, border, edge, extension, ecosystem, environment, park, site, place, landscape. =	Aristotelian categorical, natural history proposition, form of life, natural goodness, primary and secondary goodness

(continued)

Box 18.1 (continued)

	Species conservationisms	Space conservationisms	Conservationism of life forms
	population genetics ... Number, trend, abundance, distribution, biodiversity, extinction, population ...	Terrestrial and marine environments, wetlands, jungles, forests, deserts ... Territories, property, jurisdictions, sovereignty Pristine, degraded, preserved Habitat destruction, desertification, expansion of the agricultural frontier, etc. Protected area, National Park	Organism, representative, specimen, individual, particular, form, species in the sense of life form
Examples of typical propositions and logical structures	Species *originate*, *evolve* and *become extinct* Species are natural capital "Birds are beautiful" (BirdLife International) "Save the Whales"	Spaces are mapped and explored, managed, degraded or restored Ecosystems provide services The language of environmental and ecosystem ecology, landscape biology, ecological economy	*S is/has/does F* Teleological propositions of natural history (Aristotelian Categoricals) such as: ... flying squirrels have their limbs and even the base of the tail united by a broad expanse of skin ...
Evaluative language	Dependent on a variety of value systems. Predominance of use values. Intrinsic value. Good is diversity and abundance. Good is use. Useful and harmful species	A variety of value systems. Predominantly aesthetic and use values Intrinsic value Good is ecosystem service	Primary goodness is the creature's autonomous satisfaction of the needs of its form
Language of extinction	Red List language. Species "going extinct." "Sixth mass extinction." Population crash, range decline, endangered species, threatened species	Ecological impact of species extinctions on the ecosystem 'Environments' going extinct, used as an analogy with species	List of Aristotelian categoricals that no living being can satisfy, but once did
Primary value	Dependent on a plurality of value systems. The useful,	Same as for species conservationism	Natural goodness, which depends on the

(continued)

Box 18.1 (continued)

	Species conservationisms	Space conservationisms	Conservationism of life forms
	the beautiful, the marketable, the 'spiritual.' Existence, intrinsic, inherent values.		shape of a creature's life
Source of value	Derived human benefits. Evolutionary history. Life itself	Derived human benefits. Ecological functions and processes. The sustaining of life	The form-bearer relation, the here-and-now satisfaction of Aristotelian categoricals
Reasons to intervene	'Threats' of extinction. Loss of abundance. Loss of natural capital. Abuse. Compromise of aesthetic value	Loss of the 'pristine.' Compromise of ecosystem services, aesthetic or use value, etc.	Human interference with the satisfaction of life form needs, unjustified by needs of the human form of life
Interventions	Cost-benefit negotiations. Policy tools. Solutions based on spatial management (protected areas, etc.). Management plans. Adaptive management . . .	Protected areas Spatial planning models Management plans	Use of language: explicit expression of judgments (stating clearly the reason to intervene). Various approaches to express the unacceptable. (Some interventions are similar to species and space conservation, such as creating protected areas to guarantee the satisfaction of primary necessities)
Primary badness	Over-exploitation. Population decrease. Decrease in distribution. Loss of ecological function. Abuse. Failure to recognize intrinsic value	Habitat destruction or degradation	Unjustified interference with satisfaction of needs according to the Aristotelian categoricals
Meaning of "extinction"	Population size 0. "Disappearance" of the species (with all the consequences according to the valuation system)	Spaces degrade, are threatened and 'disappear,' do not go extinct. The term may be applied by proximity to species	Elimination of the possibility of any further bearer of the natural goodness of the life form

(continued)

Box 18.1 (continued)

	Species conservationisms	Space conservationisms	Conservationism of life forms
What is conservation?	Ensuring the viability of populations Maintaining population size and distribution Maintaining biodiversity Eliminate or mitigate threats Promote values	Ensuring 'healthy' ecosystems. Representativeness of ecosystems and environments in the protected areas. A percent of the planet's surface under some form of spatial protection	Represent threats with the language of categoricals. Make explicit the process that justifies interventions. Make explicit primary and secondary values in the interventions. Correct interference to natural goodness, unless justified by necessities of the human form of life. Train in the virtues related to comprehending and addressing the natural goodness of all forms of life

Box 18.2 Intervention

The corrective actions in conservation, meant to redress threats to species or places, are called "interventions." Examples are quite various and can include the following: developing a management plan, planning a communication campaign, promoting the enactment of a law, enacting the law, creating a protected area or providing scientific data for decision-making, interfering with the course of a whaling ship, combating illegal wildlife trafficking, and restoring an affected natural environment.

The pragmatist understands conservation to be primarily intervention against over-exploitation and destruction of environments, for example. To intervene implies acting on the basis of some justification for doing so. The interventions and justifications of the various practices are not always compatible. For example, defending animal rights through public demonstrations is not compatible with gathering data in support of the sustainability of trophy hunting.

The language of representation and evaluation is decisive. Each intervention responds to value judgments, and these are expressed through a variety of language-games. There are conservation practices based exclusively on secondary values for the affected living beings. But for the conservation of life forms, an intervention is an action based on the standard implicit in the language of representation-evaluation of life forms.

(continued)

Box 18.2 (continued)

General remarks about what should guide interventions

1. Before carrying out an intervention, the language-game on which it is based must be understood. Priority should be given to understanding how the language represents relevant states of affairs. The conservation of forms of life should not deviate, when justifying its interventions, from the grammar of representation-evaluation of life forms.
2. Each practice should respond to what it understands as a threat, with the destruction of habitats and over-exploitation as of primary concern. The conservation of the forms of life expresses threats always with reference to Aristotelian categoricals.
3. In practice, threats and, consequently, also interventions tend to be depersonalized. The conservationism of life forms is based on the joint evaluation of the act and the individual who carries it out.

References

1. Max, D.T. 2014. Green is good. *The New Yorker*, May 12.
2. IUCN World Commission on Protected Areas. 2004. Protected area categories. *Parks Magazine* 14 (3). https://www.iucn.org/content/protected-area-categories. Accessed March 11, 2023.
3. Gray, B.E. 2023. Yellowstone National Park Act (1872). Major Acts of Congress. *Encyclopedia.com*. https://www.encyclopedia.com/history/encyclopedias-almanacs-transcripts-and-maps/yellowstone-national-park-act-1872. Accessed March 11, 2023.
4. U.S. Department of the Interior. 2016. Secretaries Pritzker, Jewell Applaud President's Expansion of the Papahānaumokuākea Marine National Monument. *Press Release*. https://www.doi.gov/pressreleases/secretaries-pritzker-jewell-applaud-presidents-expansion-papahanaumokuakea-marine. Accessed March 11, 2023.
5. IUCN World Commission on Protected Areas. 2023. *Our Mission*. https://www.iucn.org/our-union/commissions/world-commission-protected-areas. Accessed March 11, 2023.
6. Wittgenstein, L. 2009. *Philosophical Investigations* (trans: Anscombe, G.E.M., Hacker, P.M.S. and Schulte, J. Revised fourth edition by Hacker P.M.S. and Schulte J.). West Sussex: Wiley-Blackwell.
7. Kareiva, P., and M. Marvier. 2012. What is conservation science? *Bioscience* 62 (11): 962–969.
8. Marvier, M., Kareiva, P. and Lalasz, R. 2012. Conservation in the Anthropocene: Beyond solitude and fragility. *Breakthrough Journal*, Feb 1.https://thebreakthrough.org/journal/issue-2/conservation-in-the-anthropocene. Accessed March 11, 2023.
9. Campagna, C., and D. Guevara. 2014. Conservation in No-Man's-Land. In *Keeping the Wild: Against the Domestication of Earth*, ed. G. Wuerthner, E. Crist, and T. Butler, 55–65. Washington, DC: Island Press.
10. Jacquet, J., E. Patterson, C. Brooks, and D. Ainley. 2016. 'Rational use' in Antarctic waters. *Marine Policy* 63: 28–34.
11. Batavia, C., and M. Nelson. 2016. Heroes or thieves? The ethical grounds for lingering concerns about new conservation. *Journal of Environmental Studies and Sciences* 7 (3). https://doi.org/10.1007/s13412-016-0399-0.

12. Cafaro, P., and R. Primack. 2014. Species extinction is a great moral wrong. *Biological Conservation* 170: 1–2.
13. Doak, D.F., V.J. Bakker, B.E. Goldstein, and B. Hale. 2014. What is the future of conservation? *Trends in Ecology and Evolution* 29 (2): 77–81.
14. Mccauley, D. 2006. Selling out on nature. *Nature* 443: 27–28. https://doi.org/10.1038/443027a.
15. Noss, R., R. Nash, P. Paquet, and M. Soulé. 2013. Humanity's domination of nature is part of the problem: A response to Kareiva and Marvier. *Bioscience* 63 (4): 241–242.
16. Wuerthner, G., E. Crist, and T. Butler. 2014. *Keeping the Wild: Against the Domestication of Earth*. Washington, DC: Island Press.

A Bridge from Natural Goodness to Morality 19

The main purpose of this work is to propose a practice—the conservation of the forms of life—which, we believe, would bring about a fundamental change in how conservation has been predominantly understood and practiced, in, at least, the last 50 years. It would fundamentally change the *reasons* for doing conservation, from those that operate now, especially in the conservation of life in all its great diversity of forms and the effort to protect them from the extinctions we unnecessarily threaten and provoke. The change would come from making the necessities of the life forms themselves the explicit and primary standard for justifying and motivating all practices and interventions. Not just one more practice among many others, but rather one fully grounded in, justified and guided by, the value implicit in the forms of life themselves: natural goodness. This fundamental change in practice, justification and motivation should, on other hand, seem like a homecoming to many of us, and thus, in a way, no radical change at all, but instead a return to what made so many of us conservationists and environmentalists in the first place.

We have argued that natural goodness is the primary value for conservation and that it can be expressed by categorical propositions schematized by "*S is, has, does F,*" where S = life form and F = an attribute or action or function. The logical implications of the language of the conservation of forms of life differ from those of the other language-games in conservation, so the practice requires fidelity to its grammar, without divergence from the representation-evaluation of what living things are, have, and do according to their forms. As will be developed in Chap. 23, natural goodness justifies interventions that are often quite different from any of those derived from the secondary value judgments emphasized by other practices. Explicit comparisons with other practices will be made in Chap. 24. And even where interventions may overlap, we will indicate how the standard of natural goodness provides its own unique grounds for them.

So far, we have dealt with the language of life forms and natural goodness, and its *a priori* template, which represents the mutually interdependent and necessary relation between creature and form, according to which all empirical observations of living things must be cognized. The specific shape of any given form of life is

C. Campagna, D. Guevara, *Speaking of Forms of Life*, Fascinating Life Sciences,
https://doi.org/10.1007/978-3-031-34534-0_19

understood on the basis of judgments that describe the specific needs of that form, needs whose satisfaction constitutes natural goodness. These are objective judgments, independent of cultural preferences, feelings, aesthetic values, spiritual values, etc. Moreover, natural goodness is neither intermediate to, nor dependent on, any other value when it comes to the evaluation of living things. Other values are relative to different standards, not necessarily objective or universally applicable to living things, nor connected to the needs of their forms. They are, on the contrary, often in conflict with those needs.

The justification and grounds of the practice of the conservation of life forms, and its redress of provoked extinctions, lie in our development of the theories of Thompson and Foot. It is worth emphasing: *If there is a logical grammar necessary to understanding living things, it would seem that any language meant to address the harm we do to them should be grounded in it.* We emphasize this as it is, in many ways, the central claim of our book. Alternative language, which may support a variety of conservation practices, generates evaluations that set aside, or subordinate, the standard that seems most relevant to that harm: i.e., natural badness. This consequence of the use of alternative language is most apparent when the reasoning that leads to certain interventions is made explicit. This aspect of life form conservation will be developed in Chaps. 23 and 24. When made explicit we see that the value judgments behind the common interventions in conservation are not based on the crucial relationship between life form and bearer, and the standard of natural goodness implicit in it For example, what is good for populations—as we have seen—does not necessarily satisfy standards of the natural goodness of the living things that make them up; likewise the satisfaction of aesthetic standards, for example, is not necessarily consistent with natural goodness. As a consequence, secondary values sustain the dominant conservation practices, and alienate them from the primary good, which in turn lead to interventions and other human actions that hasten extinction. And even when they prevent such actions it is not for reasons necessarily linked to the necessities of life. This tends to result in ineffective interventions, or in the establishment of 'corrective' actions that do not address the problem and that, even if they do, may be reversed at any time, when alternative opinions prevail. We have already discussed how a plurality of values confuses the reasons for action in conservation. The practice of the conservation of life forms is the only corrective reliably guided by the value in life itself.

Human Morality

Up to this point we have dealt with the notion of natural goodness without going into the main use that Foot makes of it: application to morality between humans. As we have seen, Foot considers morality to be a form of natural goodness for human beings: natural goodness of the human will and character. We want to begin, here, to investigate the relation between human acts that extinguish species and the Aristotelian categoricals that describe morality. There are many language-games in conservation, as we have already seen, which attempt to express moral judgments about human-induced extinctions. If, as Foot argues, the grammar that represents human moral goodness is the same as that which represents the natural goodness of any other life form, it seems highly pertinent to investigate the grammar of moral goodness in the context of human acts that affect non-human living things. On what natural historical judgments do moral judgments about the harm we do to other creatures rest? What is the basis for thinking there are any such judgements?

With the grammar of natural goodness, we can evaluate the effects of our actions on threatened species, according to the standards of their form, and at the same time evaluate the same actions under the standards of our own form. Any judgments we make about the acts done—e.g., that they are harmful—are necessarily related to those who carry them out. This is so because the standards of natural goodness and badness apply, in the first place and necessarily, to living *individuals*, *not* to isolated acts. In other value theories, including those behind many conservation practices, there is commonly no need to make any reference to the agent, in evaluating an isolated act. Consequentialism provides the best example where the overall 'good' state of affairs is the ultimate goal, and where cost and benefit compromises as a means to that ultimate end are permitted and even required. Moreover, there is no need to identify responsible actors, so long as the end is achieved. This way of proceeding is very influential in conservation practice. Thus, a fishing operation is halted in one place, but allowed to start up in another, for an overall 'better' effect. It is commonly not relevant to the intervention to identify the agents (companies, corporations, the ship captains, etc.) who created the need to intervene in the first place. In some instances, there may be some after-the-fact attempt to list the nations or companies that contribute most to the problem. But the tendency is not to hold actors specifically responsible when intervening.

In contrast, life form conservation makes the individual, and what it has and does, the primary focus of assessment. The focus is on, for example, whether the shark's need to move through the water with its fins is satisfied, among many other categoricals. This is not a consequentialist approach; it sets limits, based on the categoricals, to what can be done to achieve overall outcomes. The non-consequentialist approach is badly needed in conservation and, fortunately, is being pursued in some places [1].

In sum, the connection between agent and act is especially pertinent to *moral* judgments applied to the actions of human beings and their effects on other life forms—particularly, moral judgments based on the grammar of "the forms of life." But can we in fact apply moral judgments in this way?

Foot seems to think not:

> It may seem that the suggestion of a form of goodness common to all living things must carry implications about the way we should treat animals and even plants. But this is a complete misapprehension. Moral philosophy has to do with the conceptual form of certain judgements about human beings, which cover a large area of human activities. Thoughts about cruelty to animals, or about the wanton destruction of useful or beautifully ingenious living things, belong within the usual distinctions of virtues and vices [2, p. 116].

However, what Foot emphatically denies at the start of the passage, she seems to leave open as a possibility at the end. The traditional distinctions between virtue and vice are *moral* distinctions, as she herself notes when introducing her book:

> What is this book about? It is a book of moral philosophy, which means that we must discuss right and wrong and virtue and vice, the traditional subjects of moral judgement [2, p. 1].

But then questions about cruelty to animals or deliberate destruction of living things (as in provoked extinction), or any other questions having to do with our treatment of non-human living things, might be *moral* questions raised by standards of natural goodness. We can then ask, *on the basis of natural goodness*, whether cruelty to other living things, or the utter disregard for the forms of life in our provoking their extinction, is morally vicious. We can ask if compassion and respect for them are morally virtuous.

On the one hand, then, Foot seems to restrict the application of moral philosophy to human beings and their relationships with each other, so that the grammar of moral judgments about human beings would not guide our interactions to other forms of life. On the other hand, Foot's directing us to virtue and vice as a guide seems to pull us in the opposite direction, inasmuch as the traditional study of virtue and vice is focused on the morality or ethicality of our will and character ("moral" derives from the Latin translation of the Greek "ethics" and we use the terms interchangeably). The ancient roots of the tradition are in Aristotle, whose major work in ethics focuses on the virtues or excellences (*arete*) of human will and character [3].

We propose that the conservation of life forms be understood as a moral practice in this tradition. We are developing Foot's idea that moral vice is a form of natural badness or defect, and virtue a form of natural goodness, of the will [1, p. 81]. In her Postscript, Foot recounts the story of a certain man who, owning an aviary, was so struck by the beauty of a dead bird that he had the rest killed and stuffed. There are many things one might judge wrong about the man: his aesthetic tastes, his impulsiveness. But the utter disregard for the creatures themselves would certainly seem to be among them. With all due respect to her ambivalence, we suggest an extension of Foot's theory of natural goodness that allows moral evaluation of this strange man, on the basis of the same standard the theory uses to evaluate the effects on the birds: the satisfaction of the necessities of the respective forms of life. The extension we propose of Foot's theory turns on the working principle that regard for the natural goodness of other creatures is a necessity of our own form. In other words, human ethical virtue requires respect for the natural goodness of living things generally. This seems to us to be a safe assumption, one that follows from the essential human capacity to recognize value wherever we find it and to acquire some sense of its nature or importance. Perhaps foolishly trying to go where angels fear to tread, this is the bridge between natural goodness and morality in conservation that we hope to begin to build.

References

1. Batavia, C., and M. Nelson. 2016. Heroes or thieves? The ethical grounds for lingering concerns about new conservation. *Journal of Environmental Studies and Sciences* 7 (3). https://doi.org/10.1007/s13412-016-0399-0.
2. Foot, P. 2001. *Natural Goodness*. Oxford: Clarendon Press.
3. Aristotle. 2003. *Nicomachean Ethics*. The Project Gutenberg eBook 8438. https://www.gutenberg.org/files/8438/8438-h/8438-h.htm. Accessed March 11, 2023.

20 Natural Goodness Encompasses Moral Goodness

The way through to the future is by rethinking our priorities and by starting reforms- as sweeping as the moral shift that ended slavery . . .
R. Payne [1, p. 339]

In the last chapter, we have identified, even in Foot's ambivalent position, a way of possibly extending morality to our relationship to the other life forms, all based on the primary standard of natural goodness. In this chapter, we continue to clarify and develop our thought in this direction, and also offer a compelling example from the history of conservation—the *Save the Whales* (STW) movement: a, once, extraordinarily effective conservation practice strongly based on a moral imperative, grounded at least in part in natural goodness [2]. An important insight that emerges has to do with how natural goodness encompasses morality.

The STW movement of the 1960s and 1970s momentarily halted industrial-scale slaughter of some cetaceans when many species were near extinction. The unconditional, imperative language of their slogan itself reflects the movement's uncompromising aspirations: to condemn industrial whaling as evil and unacceptable. Some interventions at the time were still based on the more 'respectable' language-game of population science, articulated in the context of policy negotiations at the International Whaling Commission (IWC). It was argued that the models on which hunting quotas were based were flawed and needed to be revised. But the radically new movement also argued for interventions based on the natural history of whales, on the appreciation of what whales are, have, or do, according to their form. Scientists and naturalists made the case that whales are characterized by complex social behaviors, and that some communicate through elaborately structured 'songs.' These natural historical facts inspired and drove the movement.

The characteristics of the natural history of whales were understood as a primary value and reason for conserving them and other marine mammals. And it was on these values that expressions of repudiation and disapproval of the slaughter were

C. Campagna, D. Guevara, *Speaking of Forms of Life*, Fascinating Life Sciences,
https://doi.org/10.1007/978-3-031-34534-0_20

sustained, and interventions that succeeded in stopping the hunt for a time. In their effort to right the wrongs they perceived, STW activists undertook public demonstrations, as well as 'radical direct actions,' where, e.g., activists maneuvered their boats to directly interfere with hunting boats. The interventions were accompanied by demands for an end to whaling, demands grounded in animal welfare messages, but also in recognition of the life form needs of the species as primary. In short, the whaling industry was condemned as unacceptable because it made it impossible for whales to live as they should, to satisfy needs that naturalists were newly discovering and activists were recognizing as of primary value. There was no compromise with economic or social considerations, nor was 'tradition' cited as part of a pluralistic effort to include all points of view. The demand for negotiators was "*Save whales, not whalers!*"

One of the most influential naturalists of the STW movement, who made key contributions to our understanding of the Aristotelian categoricals describing the forms of life in need of protection, was Roger Payne. Years later, in the book *Among Whales,* Payne expresses a position on conservation, epitomized in the quote at the beginning of the chapter, but worth citing in context here:

> Band-Aids, like establishing watchdog committees, or passing new laws, won't work. The way through to the future is by rethinking our priorities and by starting reforms—as sweeping as the moral shift that ended slavery ... If what I am saying contains a great truth, how do we go about achieving such a moral shift? Again, I feel it can be started by getting people to fall in love with the wild worlds. We defend freedom with our lives; we don't defend it because it has a favorable cost-benefit ratio. We preserve it, fight for it, even die for it, because we love it so passionately that we will not live without it. I feel the same about the wild world, and I sense that I am not alone [1, p. 339].

This is an example of a moral argument of the sort whose possibility we have mentioned a few times throughout and sketched a possible bridge to in the previous chapter; it elegantly and boldly proposes a moral revolution in our treatment of other species, comparable to the abolition of slavery. The longing for radical moral change, understood as a profound revolution in human values in relation to the other life forms, is deep, for anyone concerned with conservation of life forms. But it must be understood that such profound reforms face a fundamental, conceptual problem that abolitionist claims do not. The ideals for which a virtuous human being would be willing to die are expressed in familiar language-games, for example, the language of liberty and equality. But as we have discussed many times already, the extension of such ideals, in our relationship to other forms of life, invites confusion.

In contrast, the language of natural goodness is a language that *encompasses moral language* as part of the natural historical description of *human* natural goodness, just as it also contains the natural historical language that shows us that whales communicate by song, etc. Even if we assume (as in any case seems evident) that some species possess some human-like attributes, the theory of natural goodness indicates that there are innumerable other reasons for valuing them, more important and proper to those life forms than resemblance to humans. The theory is grounded in a logical grammar of the good, applicable to all living things, on the basis of the relevant Aristotelian categoricals. What might be the categoricals, if any, that represent the necessities proper to the relationship between human beings and other forms of life? Are there any that could amount to an ethic leading to the radical change we seek, one that overcomes the dominant homocentric, instrumentalist ethics based on a plurality of values secondary to, and often in conflict with, natural goodness?

How Would a Morality Based on Natural Goodness Make a Difference to Conservation Practice?

In this chapter, we will be developing our response to this question and related questions just raised in the last section: for example, what might we gain that is new from the practice of the conservation of life forms, as compared to the dominant alternative practices (some of which do claim that human behavior towards the other life forms is wrong morally)? How *would* such a practice be implemented or justified? For the moment, the following can be noted:

1. Human beings understand and value what it is to satisfy the needs of their form, and they know, at least implicitly, how to represent-evaluate living beings on the basis of categoricals that express the needs of life forms in general. Presumably, a virtuous person would reflect not simply on how she could respect those needs in her own life form but also those of the others, since they are instances of the same general value. These reflections are already ethical reflections on virtue and a basis for developing it in relation to non-human life, on one and the same standard.
2. Reflecting on the natural value instantiated by individuals of all the various forms of life, is therefore part of virtue. As was suggested in the last chapter, where we considered the vice of cruelty and virtue of compassion, it seems reasonable to consider extending the scope of what is required of us by traditional virtue theory to the other life form, but always based on and guided by natural goodness in each form and, of course, in what ours requires in light of that of the others.

Regarding the practice of conservation of life forms, the most obvious place to begin exploring the relevant virtues or vices is with those that all humans already understand, such as cruelty, or lack of empathy for fellow human beings, or the deep disregard for the natural goodness of our life form that that involves. This follows Foot's suggestion, found in her story of the bird collector who was "struck one day by the beautiful appearance of a dead bird that he straight away had the rest killed and stuffed." Foot continues with this remark [3, p. 116]: "Hardly a crime! And yet there was something wrong with that man." What is wrong with that man is to be judged by the same standard that shows what is wrong with the industrial whalers, the traffickers of wildlife, etc., the only standard that can assess the meaning and measure of their viciousness toward other life forms. And we will have to consider what is wrong with us all too that we routinely allow or contribute to the unnecessary annihilation of species. Is there any reason to think these are not moral failings, unjustifiable and unacceptable wrongs, to be resisted and corrected accordingly?

Dominant conservation practice has lost sight of this perspective, in sharp contrast to the Save the Whales movement. What would it mean to return and stick to it as the primary basis for "doing conservation"? These are ethical questions, questions primarily focused on commitment to the implications of our own natural goodness, in recognition of the natural goodness of other living things. The difference it would make would be seen, first and foremost, in a person's character and will, which is where every enduring change for the better begins or ends, and from which the desired actions flow, as it has been with rejection of racism and other evils, to the extent that we have overcome them. In this regard, Payne's eloquent plea, for love of nature as a moral basis for preserving it, is necessarily limited. We cannot love all other human beings, let alone all other forms of life and their individual bearers. What we have fallen in love with and are willing to die for in our relationship to fellow human beings are ideals and principles like equality and freedom. When we have learned proper respect and awe for these ideals, we treat others—whether loved ones, or strangers, or even enemies—on their basis. What Foot's theory teaches is that these values are ultimately to be understood in terms of the standard of our natural goodness: in terms of what we are, have, and do as human beings. Natural goodness is primary and encompassing of moral goodness. The view advanced in this essay, daring to go beyond Foot's explicit theory, urges us to recognize human natural goodness as involving respect for what all other living things are, have, and do, according to their own forms of life, which is just another specific instance of natural goodness.

At this altitude, the theory cannot say much about what might be more specifically required of us in our treatment of the other life forms. At this level, Foot is right to question the kind of application of her theory we are entertaining. In order to keep the application as uncontroversial as possible at first, we have concentrated on routine and avoidable provoked extinction, and the egregious disregard of other creatures shown by certain practices like shark finning, trophy hunting, and clearcutting. There will be many difficult cases to discuss and evaluate in a complete theory, but surely these are morally vicious and unacceptable, if anything is on the basis of natural goodness. And the conservation of life forms must redress and resist them as such, without compromise with any other supposed value.

Box 20.1 The Success of "*Save the Whales*"

1. The *Save the Whales* movement illustrates the notion of conservation of life forms in the following ways [2]:
2. *The form of life and individual creature bearing the form were simultaneous targets of the intervention*. Not all actions taken by the participants of the movement were justified solely on the basis of life form needs, but those

(continued)

Box 20.1 (continued)

needs were central to the reasons for intervention. The interventions were executed often primarily on behalf of individual creatures, justified solely by concern to protect satisfaction of the needs of their form. This is life form conservation practice.

3. *Each individual creature mattered and not just because of its contribution to the size of a population.* The movement was about saving species, but via preventing the killing of individuals. Individuals were not protected exclusively as a means to the ultimate end of protecting the population they made up; they were protected at least in part for the sake of their autonomous satisfaction of the needs of their forms. Direct intervention was intended to prevent the death of individuals, for that reason The language of the interventions was applied to the individual, just as the language of natural goodness requires.
4. *Some interventions were grounded in natural good and evil.* As implied by the above, principal interventions were motivated by the natural goodness of non-human life forms, and the natural badness of the effects of certain human actions on them. Some interventions were reminiscent of the human rights struggles of the mid-twentieth century, and may suggest the language of rights as a basis. However, the movement's sensitivity to the emerging natural history of whales and other creatures illustrates that even 'radical direct action' is consistent with a practice based on the theory of natural goodness.
5. *Practice was sensitive to the logic of the language of life forms.* Some secondary values that lead to confusion of standards tended to be avoided or were rejected, such as those that place economic values over the flourishing of individuals, or use values over the satisfaction of primary natural needs.
6. *The standards for conservation interventions were expressed by the same grammar for all forms: namely, that which represents the needs of the forms.* The focus was on "the whale" as a generic or kind, which encompassed many species under the general life forms. This reflects the grammar of generics peculiar to life forms and their natural goodness. As in life form conservation, judging some species 'better' than others, or as more 'charismatic,' 'sensitive,' 'attractive,' or as 'harmful,' and 'dangerous' to us, was understood to be secondary or derivative, and not justified by the primary standard, generically the same for all.

Box 20.2 The Demise of "*Save the Whales*"

The practice of life form conservation condemns industrial whaling primarily on the basis of two things: the effects on the individuals of the form and the absence of any need of the human life form to justify the activity. Industrial whaling involves killing individuals who are splendidly satisfying the needs of their form.

The *Save the Whales* movement prevented the hunting and killing of many whales, but did not install permanent values based on the language of natural history and its Aristotelian categoricals. Use values re-emerged, neutralizing the anti-whaling attitude or confining its language to that of of 'populations' and 'sustainability'. After a moment of splendor, the grammar of natural goodness faded into obscurity, supplanted by the language of secondary values, grounded predominantly in economic costs and benefits, or political expediency, cultural traditions, and the sovereign decisions of nations. At the same time, scientific development expanded the frontiers of the knowledge of life forms, without each discovery being incorporated into the language of management.

Today, arguments in favor of whale killing, articulated in terms distant from the language of the natural history of animals, are advanced by interest groups that tend to belong to welfare economies, with human populations that do not require the products of the whaling industry. These groups defend their interests against the values of alternative groups, a conflict that is settled in the arena of political force rather than ethical argument. *Save the Whales* was an impressively distinctive movement, an ethical movement, which showed what can be accomplished on the values we espouse in this essay. At the same time, it also shows that the virtues needed to sustain lasting change are not installed simply through having grasped the right values and language. This highlights the need for virtue: an enduring character to reliably act well, with good judgment, across many often challenging circumstances. The return to instrumental valuation was the consequence of the sum of small 'victories' at all levels, particularly at the International Whaling Commission. Similarly, discussions about biodiversity or climate change are fully controlled by language-games that expel alternative valuations as wrong, naive, or, paradoxically, even anti-social.

Still, the practice of conservation of life forms must maintain its position by not compromising its language or the standard implicit in its grammar. It understands that the use of language is itself an action of conservation, an intervention. The use of language is a fundamental human activity and the analogy to games brings out how it is interwoven with the rest of human life, with everything we think and do. But to understand the practice of life form conservationism requires putting a particular grammar into action, applying language as an intervention that triggers other interventions in the style of the

(continued)

Box 20.2 (continued)
Save the Whales movement. Some specific practical application of this aspiration will be discussed in the final chapters.

References

1. Payne, R.S. 1995. *Among Whales*. Charles Scribner's Sons, Simon & Schuster Inc.
2. Campagna, C., and D. Guevara. 2022. "Save the Whales" for Their Natural Goodness. In *Marine Mammals: The Evolving Human Factor*, ed. G. Notarbartolo di Sciara and B. Würsig, 397–424. Cham: Springer. https://doi.org/10.1007/978-3-030-98100-6_13.
3. Foot, P. 2001. *Natural Goodness*. Oxford: Clarendon Press.

Agent, Action, and Modalities of Action 21

There is nothing astonishing about certain concepts only being applicable to a being that e.g., possesses a language.
L. Wittgenstein [1, §520 p. 90]

Summary of Some Key Terms and Distinctions

The passenger pigeon was treated like a 'plague,' killed to extinction. True enough, but expressed without mentioning a responsible agent, submerging the fact that there were individual human beings responsible, with names, identities, life histories, perspectives. The condemnation of Walter Palmer or King Juan Carlos of Spain, as trophy hunters, explicitly targeted specific individuals. But the common practice in conservation is to speak of and assess provoked extinctions in terms that cloak our responsibility for them [2], on top of assessing them according to secondary values. The common talk is, for example, of "overexploitation and its effect on sustainability." Even when human individuals are the subject of an assessment, they are referred to in a general and diffuse manner: as, for example, "consumers of ivory parts," "wildlife traffickers," "Japanese or Norwegian (Icelandic, etc.) whalers," "illegal, unreported, unregulated fisheries." This has an effect on the urgency and nature of our interventions.

As discussed in previous chapters, the practice of life-form conservationism, as we would develop and defend it, does not evaluate acts independently of the responsible agent; it evaluates the trophy hunt and the hunter as an interdependent unit. It assesses wrong or right action, good or bad, in relation to whether it proceeds from virtue or vice in the agent. The assessment unifies the various relevant parts in a specific judgment, in accordance with the one generic standard, natural goodness, reflected in all life forms and their relation to their bearers. The judgment evaluates the affected creature in terms of *its* natural goodness or badness. A harpooned whale is judged to be defective, injured. The specific evaluation of the human act and agent that brought it about is evaluated according to the same general

C. Campagna, D. Guevara, *Speaking of Forms of Life*, Fascinating Life Sciences,
https://doi.org/10.1007/978-3-031-34534-0_21

standard, the natural goodness or badness of the agent and action, as determined by the relevant life form, our own. Of course, as has been emphasized throughout, this is the standard implicit in the Aristotelian categoricals which describe the respective forms. While it is easier to understand and redress threats that impact some familiar life forms, the same process of evaluation and interventions based on it apply in the same way to any life form. There may be less familiarity with the natural goodnes of bacteria, fungi, or unicellular algae than with that of some mammals, for example, but the evaluative language does not change.

Common terms of evaluation in conservation, like "harm" and "threat," are used in a variety of language-games. In life form conservationism, we propose that *harms* be understood as those brought about by human acts that cause natural badness in individuals. We use "harm" as involving a "wrong" only when the harm is not justified by the Aristotelian necessities of the form of life. Thus, hunter-gatherers kill for food. The killed representatives of forms of life are dead due, of course, to having suffered injuries, and thus defects or natural badness, but they have not been harmed in the sense of "wronged," the hunter has done nothing wrong. We speak of *threats* to species, generally, rather than to individuals. The threats of primary concern are those of provoked extinction (which it is doubtful are ever justified). Thus, harpooning a whale the way industrial fishing does obviously creates defect by interfering profoundly with the animal's vital functions, and also harms and does wrong since not justified by our life form needs. The harm may or may not threaten the species with extinction.

For a couple of centuries, industrial whaling posed a systematic threat to the various life forms targeted by the practice, and could not be justified by our life form needs. In conservation based on the language of populations, the harm that may be caused to a particular whale may not be relevant, and in any case does not translate into 'threat' until the number of whales killed compromises the abundance of the population. This might seem to be a use of "threat" similar to that emphasized by life form conservation, in that it looks like the term applies, in both cases, only to large-scale killings (the form of life encompassing all representatives). However, in the conservation of life forms, the harm to any individual is in a way also always a threat to the species (though not necessarily of extinction). It is so because of the logically interdependent unity of form and bearer; in this case it proceeds from the agent's disregard for the requirements of the form, and can be compared to how we judge human-to-human harms that are not in self-defense or otherwise justified: such harms to any one individual are a threat to us all, in their disregard for humanity. Because of the logical unity that binds individual to the form it bears, the disregard for the individual is necessarily a disregard for its form of life.

Modality of Satisfaction of Life Form Needs

It is an important and distinctive feature of the conservation of the forms of life that every harm or threat is potentially a reason for intervention or redress, directed at the responsible agent. As just noted, not all assessments of harm, in the general practice of conservation, justify the conclusion that there is a wrong or a threat, and a need to intervene. But in the practice of life form conservation unjustified harm justifies intervention. As noted before, a predator kills prey as part of the natural goodness of its form of life. In the process, and in an ordinary sense, the predator obviously *harms* the prey. According to our suggested terminology, it causes defect but it is not a wrong or harm in that sense. In the same way, a parasite sickens its host by satisfying Aristotelian categoricals of its form. The host shows defect, but, in our sense, the parasite does not do anything wrong; what it does is natural goodness. Humans too are justified by life form needs when they must kill to protect themselves from a hostile environment, from parasites and predators, etc. Nothing wrong with that. The need to intervene according to conservation of life forms only arises when actions and agents harm and threaten living beings, without being justified by the needs of their own form.

Because there often are so many ways of satisfying the necessities of a life form, one might think anything could be justified; any action could be linked some way to a categorical. Humans need recreation. Therefore, trophy hunting may be justifiable as recreation. The fallacy here lies in treating all modalities of the satisfaction of needs as themselves necessary. One could indeed justify anything this way: even murder for fun, for example. But the modality matters. The fins cut from the shark are used in a meal. But it is not *necessary* for us to eat this way for flourishing; he that does not eat shark fin soup is not defective. The *modality* used to meet the need to eat something tasty, nutritious, and available is not itself a necessity. Aristotelian categoricals about our need to kill to eat do not, by necessary design, differentiate between the various ways of doing it: trapping, poisoning, shooting, beating, torture, or wasteful means. But it is obvious how some ways of satisfying it do not involve the necessity of the categorical: for example, in industrial fishing a sometimes very significant proportion of a catch is discarded, thrown back into the sea, dead or badly harmed, because it has no commercial value. Thus, industrial fishing does not in general satisfy primary necessities just because some caught fish are consumed. The human being does not require industrial fishing for flourishing or survival. And, in the case of industrial whaling, the main purpose is not even to satisfy food needs. The purpose is to maintain an economic activity that secures political power and interests over key places in the ocean. We must question any activity that drives a life form to extinction and especially one that does so just because the market pays a high price to consume it. Consequently, killing for food or defense is human natural goodness only for some modalities of the act. Talk of human 'needs' with no further qualifications, meant to justify the unjustifiable, is based on a grammar not connected to the real needs of our form.

Human 'Exceptionality'

The manner in which a need is satisfied is in general only relevant when the act is carried out by a human being. The mode of killing, for example, of a group of orcas attacking a sea lion is not to be evaluated according to whether they kill expeditiously, or avoided unnecessary suffering or pain in the prey. Orcas hold the prey, take it in their mouth without killing it and release it, submerge it, and finally kill it. These behaviors are not based on choices like our own, but are simply the way the predators satisfy Aristotelian categoricals. In the same way, the natural goodness of a parasite is not compromised because of the particular way it makes the host sick and defective; rather it is judged by whether it meets the needs of its own form, which does not involve actions that can be assessed at permissible or impermissible. We might echo Wittgenstein's reminder, in the header quote, with our own: There is nothing surprising about certain concepts applying only to a form of life that, e.g., has rational agency.

True Needs

Life form conservation faces the difficulty of having to install its practice in the face of alternative practices that rely on entrenched and familiar language-games, all played mainly with secondary values. Some of these language-games attempt to ground practices based on human 'needs' defined by standards unrelated to the life form, as we have been noting. Many false dilemma's, supposed conflicts between our good and that of other life forms, get generated this way: "It's either our jobs and dignity or else conservation of life forms. You can't have both." The dilemma is forced on us by a false conception of our needs that have to do with avoidable behaviors, rather than the necessities of our forms or the necessary modalities of their satisfaction. This leads to the false perception that those who make natural goodness their primary value are naive or indifferent to human needs. The mistaken perception is a result of judgments made according to a language-game of secondary standards, considered superior to that of the satisfaction of the necessities of a life form.

All institutional or cultural evils that have ever existed have generated economic resources upon which we have come to depend, and by which their proponents have tried to justify the evil as necessary. The slavery industry is one among other pointed examples. There are hard ethical decisions about how to dismantle an evil industry that we many depend on in some way, but the language should at least be kept clear: a true "need" is only one as determined by the form of life. And not just any way of satisfying a need is permissible. The fact that slavery provides jobs is no justification for it. The mode of satisfaction of a true need must itself be consistent with the needs of life forms.

References

1. Wittgenstein, L. 1981. *Zettel*. (trans: Anscombe, G.E.M. Edited by Anscombe, G.E.M. and von Wright, G. H.). Oxford: Blackwell Publishers Ltd. https://www.informationphilosopher.com/tractatus/Collected_Works.pdf. Accessed March 11, 2023.
2. Cafaro, P. 2015. Three ways to think about the sixth mass extinction. *Biological Conservation* 192: 387–393.

22 Rationality and the Good

What I want to stress at this point is that in my account of the relation between goodness of choice and practical rationality it is the former that is primary. I want to say, baldly, that there is no criterion for practical rationality that is not derived from that of goodness of the will.
P. Foot [1 , p. 11]

We have seen that conservationists have at times condemned as immoral the human killings of other creatures, particularly those that lead to provoked extinction. But they may have done so without due regard for the implications of the language-games used to extend talk of "morality" in this way. Moreover, there is virtually no concern to try to ground this talk in the grammar of the language of life forms. Foot, as noted (Chap. 19), expresses ambivalent doubts about using that grammar to derive any implications about how the other creatures should be treated by humans, saying flatly that it is a mistake to think there are any such implications, while, on the other hand, recommending the study of virtue ethics as the place to explore answers to such questions. Other authors have explored the quality of these virtues (see references [1–3, 5–7] of Preface for Conservationists).

Yet, regarding the story of the man with an aviary that ordered all birds killed and stuffed, after being struck by the beauty of a dead one (Chap. 19), Foot says: "Hardly a crime! And yet there was *something wrong* with that man" [1, p. 116; her Italics]. It seems to us that limiting the judgment to "something wrong" is insufficient when human behavior leads to an extinction, like that which led to the extinction of the passenger pigeon, or to near extinction, like that of the dolphin of the Gulf of California. We have therefore proposed that the conservation of life forms be understood and developed as a virtue ethic, based on the standard which is generically the same for all creatures, human and non-human: namely natural goodness, i.e., a creature's autonomous satisfaction of the necessities of its form. We have cited the *Save the Whales* movement as an example of the power that comes from morally motivated commitment to conservation of life forms, and uninhibited use of moral

C. Campagna, D. Guevara, *Speaking of Forms of Life*, Fascinating Life Sciences,
https://doi.org/10.1007/978-3-031-34534-0_22

imperatives (Boxes 20.1 and 20.2). We have observed that, historically, enduring change for the better, especially where great evil has been overcome, lies in moral change. Other means, including legal, of redressing the worst tendencies in society only last, or are respected, when individuals have taken the underlying values and principles to heart, as Roger Payne suggests in his elegant plea.

However, it must be noted that what commonly counts as moral is influenced by philosophical and religious traditions which use the concept in a sense close to the language of obligation, moral law, or divine commandments. The critical limitation of this language, the language of what we 'owe' to one another, or to the state, or to God, is that it only makes sense in human life, or the lives, if any, of creatures like us with the same kind of need for social cooperation and fidelity, and the ability to make contracts, promises, and so on. Traditional theological or religious applications take this beyond the human, but in the wrong direction: to God and heavenly things, away from natural life forms on earth. All of these language-games have implications that can conflict with what is necessary for natural life forms.

Foot's principal thesis, we have seen, is that natural goodness is the primary good and that morality is only one kind of natural goodness: namely, the natural goodness of human will and character, i.e., ethical goodness. Natural goodness is the more fundamental notion, subsuming morality under it. We have proposed as a working hypothesis that it is part of this human natural goodness to recognize natural goodness in the other life forms and to see it as a basis for how we should treat them, how we should judge what is permissible or impermissible in our treatment of, and effects on, them—particularly in case of what provokes their extinction. And we have emphasized that such judgments necessarily involve an assessment of both the agent and the actions he is responsible for. Here there are always two things to keep in mind: the character and will from which the action flows, and the effects on the other life forms. What Foot and Thompson have shown, by attention to logical grammar, is that the general standard for assessing the action is the same for assessing its effects; it is the standard implicit in the grammar of the natural history of any life form.

Practical Reason

The natural history of the human life form, as we saw in the previous chapter, introduces elements that make the statement of Aristotelian categoricals much more complex and challenging in certain respects not typical of other life forms. This becomes clear when one considers that what we do, have, and are, according to our life form, depends crucially on our rational agency or practical reason. How should we understand the role of practical reason and the difference it makes to our life form? This is the focus of this chapter.

The term "practical reason" in traditional philosophy, going back to the ancients, is used in ways that are not common today. Today, the term tends to connote or imply the idea of *instrumental* rationality—reasoning about the best, most efficient, means to an end—the epitome of "practicality" in certain circles. This is the concept of practical rationality most common to conservation. On this concept, our reason cannot help us set the ends we are to follow; it does not tell us what is good to pursue or not. Desire, sentiment, tradition, culture, taste, whim, and many 'non-rational' or even irrational inclinations do that. This is in contrast to the older, more traditional view, where reason does not simply serve our inclinations but also directs us toward certain ends, rather than others, because they are good or better ends. On Foot's view, in fact, this is the primary role of reason when it comes to questions of how to live, or of what we should do, how we should be. This position is a consequence of her theory of the nature of our life form, as described by the Aristotelian categoricals pertaining to our rational agency, our character and action. These are the categoricals that describe what it is to be a good human being, by describing what it takes to satisfy those necessities of our form of life having to do with the fact that we must reflect and deliberate upon what to do, what to pursue, given what we are. We do not always have the benefits of being on the rails of instinct or automatic behavior; our natural goodness necessarily involves our reflecting on, adopting, and being disposed to follow, principles of choice and action that involve a kind of cooperation without which we could hardly survive. This is the backbone of morality, the natural goodness peculiar to our species. It introduces mind-boggling flexibility and, thus, complexity, into the specifics of how we may satisfy the categoricals of our form.

The principles of justice, fidelity, honesty, etc., are central here because otherwise the cooperation we need to live would depend entirely on power or affection or luck: a life according to which we cannot flourish. The relevant categoricals of will and character define what the natural good is for us in respect of this peculiar feature of our form of life and Foot is saying, in the header quote, that what it is for us to be practically rational requires conformity to the natural good. Natural goodness is once again primary. Unless our ends are good (according to how our form determines goodness) our practical rationality is not good. We can be geniuses at adapting means to our ends, but if our ends are not good, do not contribute to satisfying necessities of the human form, then what does it matter? If reason is only instrumental, then how could it be important or valuable? This is a position Foot spent much of her later career trying to defend. The position is the one Foot expresses in the header quote: it is a controversial position and this is not the place to try to defend it. But the view is worthy of consideration, though largely neglected today. It is a view of practical reason grounded in the concept of natural goodness. On this view, morality is a form of practical rationality, which in turn is the form of natural goodness for us.

The conservation of life forms therefore condemns, as irrational, the pursuit of ends in conflict with natural goodness; our own natural goodness or that of the other life forms (except when it is in conflict with our own). "Practical irrationality" in the language-game of the forms of life means "rationally defective," or "unjustified in respect of life form categoricals" for a rational agent. In the theory of natural goodness, as we are developing it, there is nothing rational or reasonable about provoked extinctions, nor about shark finning, trophy hunting, fishing discards, or industrial whaling, given the egregious interference with natural goodness, without justification in human life form needs. They must be opposed and forbidden accordingly. And eventually, as this view is internalized by society, provoked extinction, etc., may become as shameful, reprehensible, and intolerable as other egregious violations of human natural goodness. In any case, it is for these reasons that life form conservationists are justified to condemn, oppose, and intervene against the violations as irrational.

This account does not project human moral values onto the rest of life, but instead grounds all value in a more basic kind of standard, common to all life forms. In the remainder of this essay, we will continue to develop our argument that it is a human virtue to base our relationship to other life forms on the standard implicit in the language of life forms. It is a defect, a vice, a failure of practical reason, for human beings to treat the satisfaction of the necessities of their own form as a primary value, but not those of other creatures and, instead, to subject these creatures to values secondary to their natural goodness. It fails to respect the logically interconnected relationship between creature and form, described by the language of life forms and that makes it possible for us to experience and identify life in the first place.

Reference

1. Foot, P. 2001. *Natural Goodness*. Oxford: Clarendon Press.

Practicing the Conservation of Life Forms 23

Language is the house of Being. In its home man dwells. Those who think and those who create with words are the guardians of this home.
M. Heidegger [1, p. 217]
What is called a justification here? - How is the word "justification" used? Describe language-games! From these you will also be able to see the importance of being justified.
L. Wittgenstein [2, §486]

The conservation of life forms is based on a language game that is set in motion with the ultimate goal of redressing what humans do to provoke extinctions. Humans routinely and needlessly interfere with countless creatures autonomously satisfying the needs of their respective life forms. These acts are the major drivers of provoked extinction; they threaten the extinction of species at an unprecedented rate. With the goal of intervention and prevention, the practice of life form conservation deals with each extinction threat in four stages (see also Box 1.3):

1. Representation of life forms by Aristotelian categoricals
2. Explicit evaluation of the agent-act according to the standard implicit in the categoricals
3. Justification of any intervention, by making explicit the reasons for undertaking it, with reference to the standard in Aristotelian categoricals
4. Intervention based on the above evaluation and justification

The first step describes the necessities which help us determine whether and how the life form is threatened of the threatened form. The next identifies the state of affairs that interferes with the satisfaction of these necessities. If the interference is due to human acts, not justified by the necessities of the human form, then this grounds reasons to intervene in order to eliminate or, at least, mitigate the threat. In the conservation of life forms the reason to intervene must be made explicit. The

C. Campagna, D. Guevara, *Speaking of Forms of Life*, Fascinating Life Sciences,
https://doi.org/10.1007/978-3-031-34534-0_23

whole process is subject to the logic of the language of the representation-evaluation of life form and creature that bears it. In what follows, we describe more specifically the characteristics of each stage, using concrete examples relevant to the conservation of life forms.

The Stages of the Practice

1. *Representation*

   It all begins with the representation of the relevant life forms, and, thus, with the Aristotelian categoricals most pertinent to the threat of extinction. It is with these natural historical propositions that the language of life forms is installed, a language that establishes the logical boundaries of the three stages that follow. For example, in the evaluation of shark finning, one starts with expressions such as "sharks have fins for the purpose of swimming." Others will also be relevant. For example, natural historical propositions that describe the role of blood and circulation in the shark, among others, apply. From these we derive that sharks that have had their fins removed bleed out and slowly die. These will seem like obvious judgments to most readers, since they already know something about sharks. But the obviousness just reflects the egregiousness of the interference with the necessities of the form. It is not the isolated fact that works here; it is the form-representative relation: empirical reality placed in the "wider context" of the form (Chap. 10).

   The starting point and limits of the language constrain the practice which emerges, though the nature of the practice might not be apparent at first. An analogy might be helpful here to explain the importance of this. If a series of central pieces of a jigsaw puzzle are thrown on a table, the final figure may not be visible, but the colors and shapes of these pieces guide and constrain the way the other pieces fit together to form the final picture. Or, likewise, the first movements in Pina Bausch's choreography ("*Tied down*"), described in Chap. 2, are determinant for the dance. If the first steps lead immediately to the limit imposed by the rope, that limit will dominate the choreography. On the other hand, if the initial steps are not eventually limited by where the rope ends, the choreography could continue outside of the limits, expressing a different story; just as the puzzle would be a different puzzle if not constrained by the initial pieces. The first steps to intervene in nature begin with the language conservation chooses to use to describe the state of affairs it wishes to address, a step as critical as the first few steps of a dance or the laying down of the first pieces of a puzzle.

   Like the competing language-games of conservation and environmental practices that dominate today, the language-game of life forms develops as each user of the language contributes his or her own 'pieces' or 'moves' in the game, and a more and more complete picture begins to emerge and become entrenched, and to seem inevitable, as has happened with the language-game of sustainable development. In this way, relatively straightforward and recognizable concepts and phrases install a system of standards and values for a practice. For perhaps the most dominant and prestigious environmentalism, the starting point draws from familiar concepts in economics. The use of expressions like "ecosystem services" installs, user by user, a system of values instrumental to human development. The starting point guides and constrains everything afterward. For life forms, the point of departure is the language of natural history, whose grammar and logic we all begin to acquire competence in from an early age (e.g., in the language of life cycles and natural documentaries that we learn in primary school), even if not aware of it as such. The logic of the language of the point of departure also constrains which objects a practice of conservation targets. For the conservation of life forms, the phrase "climate change," for example, does not pick out a conservation target, unless converted into the language of the representation of life, so that the relevant state of affairs can be assessed, and an intervention deployed, according to how climate change affects the Aristotelian necessities of life forms [3].

2. *Evaluation*
The first stage lays the foundations for the next: the evaluation of the state of affairs threatening life forms and, in the extreme, their provoked extinction. Sharks whose fins have been cut off, for example, show obviously serious defects in the satisfaction of primary needs. This is an evaluative judgment based on the representation of the shark's form of life; it is an example of the representation-evaluation character of life form judgments and it can be expressed in various ways, all of which express the defect or natural badness in one way or the other. The evaluative description extends to the activity of the responsible agent. The shark finner causes natural badness in the shark, of a particularly cruel type, inasmuch as the shark drowns, thrashing around in the water. Done without justification in human life form needs, finning is especially vicious in its egregious disregard for what the shark is, has, and does. The evaluation of the agent, act, and effect on the animal are through and through one of natural badness. Intervention is grounded in this assessment, and has this natural badness as its target.

This is a point worth emphasizing, one that marks the fundamental difference with other practices. Other conservation or environmentalist practices will also condemn and intervene against shark finning (or trophy hunting, etc.), condemning it as cruel and vicious, morally wrong, even 'unjust,' and a violation of the 'rights' of the animal, or as cruel because of the pain and suffering. As we have argued at length, the crucial difference is that only the language of life forms, and the standard of natural goodness and badness implicit in it, can ground, guide, and justify conservation practice, if what we primarily care about is what is good for living things according to the standard of goodness set by their form, which after all is the standard that determines what they are, have, and do to live. Of course, the concern must include care for our own life form: i.e., for what is primary goodness for human beings, which includes virtue and moral ideals that cannot be reduced to instrumental or economic or aesthetic value, etc. This, as we have also argued, is what has been lost in the play of all the other language-games where the focus is on secondary values, values often even inconsistent with natural goodness, or inconsistent or incommensurable with each other, as in many kinds of pluralism (see Chap. 16, above). The language of life forms is a language-game that breaks the spell, the bewitchment, of these others, by keeping clear the logical grammar of the representation of life, and the standard it implicitly contains for determining life form needs, natural goodness. It brings us back to why and what we meant to be doing in conserving the forms of life in the first place. It clarifies and makes compelling the primary value grounding respect for life forms.

3. *Explicit justification of the reason for intervention*
In what we have described, we find ample reason to intervene as life form conservationists. We get a sense of the many reasons for intervening insofar as we understand the primary needs of other life forms. In life form conservation, these reasons should be as explicit as possible, or ready to be made explicit when needed. This is in contrast to how conservation practices typically operate today.

We must make it explicit why we interfere with the needs of other life forms, whenever it is claimed to be necessary to do so. This itself is part of what it means to respect the other creatures. The imperative to be explicit exposes the avoidable and indulgent damage we do to some living things in order to satisfy extravagant tastes for example, or where what we kill is mostly discarded and never consumed, and where the mode of killing is often cruel, and profoundly indifferent to the natural good of the life form and a driver of provoked extinction.

It might be objected that some humble fishermen and their families depend on such practices (e.g., the shark finning industry)—unlike the corporations and nation states whose profits soar well beyond any necessity—perhaps as the only reliable employment available to them. Certainly, the corporations and other powers should be the focus of any intervention; at the same time, we must confront difficult questions about individual participation in an evil. We must ask how active participation in cruel and egregious violation of a life could ever be justified as a mode of 'satisfying' the necessities of our life form. Logically or grammatically, the case forces us to clarify the language-game of need or necessity. The primary necessity here is ethical and the rejection of evil, wickedness, vice, cruelty, malice, indifference to the good. Philosophers, novelists like Dostoevsky, if not life's tragedies alone, are famous for concocting all sorts of hard cases where it might seem that we cannot avoid evil, as can be illustrated indefinitely with 'Trolley Car' cases, like those invented by Foot and elaborated by many others [4, 5]. The theory of natural goodness does not immediately resolve all the hard issues around such cases (if indeed they can all be resolved), but it does provide the reasons, the principle standard, guiding and motivating us as we work out the details of natural goodness, for day-to-day life and practice. We also draw strength and clarity from doing conservation for the right reasons, reasons that likely made us want to be conservationists in the first place.

4. *Intervention*

In life form conservation, establishing the connection between Aristotelian categoricals and the justifications for intervention is itself an intervention, a corrective action, since it unmasks and makes explicit the reasons for the intervention. This is a corrective to what usually happens in other practices, where the reasons and motivations are masked or taken for granted. Making everything explicit intervenes against the infiltration of other language-games, and the standards and values that influence competing interventions in a practice. Careful attention to the logical grammar of categoricals prevents sliding into practices that focus on secondary values, like those having to do with use value. In this way, "sharks have fins for swimming" and "human beings understand the life form need of sharks to swim" express the principal reasons for intervening against shark finning, and for the human being involved to change their behavior, to condemn it, and so on.

This conception of intervention and practice belongs to the main practical theme of our essay. Conservation practice has lost sight of its language-games and, thereby, of its primary values and goals. Its interventions are all affected by this. And there will be no proper or effective interventions for the conservation of life forms until we intervene to correct this inattention to language. The solution begins with making the language-games explicit, including the reasons for one intervention or another, and the values that supposedly justify it.

The planning phase of an intervention is not an intervention *per se* but a part of any intervention, yet the whole process is irrelevant and ineffectual if, especially during planning, reasons and values are not made explicit. For example, it might be that a practice promotes the creation of a protected space on the basis of instrumental reasons for doing so (to raise money for other projects), even though the original and presumed primary motivation was aesthetic: e.g., to protect it from a threat that compromises the beauty of a landscape. But the protected space winds up being administered according to use values. Thus, the exploitation of resources for use becomes in fact more important to the actual day-to-day management of the area than avowed aesthetic values in the conservation of the space. This sort of confusion is endemic to environmentalism and conservation. It even leads to changes in the original type of intervention: for example, from the creation of a protected area to a change in management tools, or from banning certain activities to negotiating a compensation on a different front, as with "carbon offsets." This is prevented in the practice of life forms conservation insofar as everything is anchored in the rigorous and systematic logical grammar of Aristotelian categoricals, which if kept explicit or near at hand guards against straying and confusion, the way well-defined and logically coherent first principles do in the application of a theory. They continually illuminate the reason for being in the practice, keeping the language-game consistent with fundamental values, hindering the confusion of competing ones getting in the mix.

Precision in the Language: Thick and Thin

The first step in the practice of conservation of life forms is therefore one of attention to and use of a certain language, and the first intervention is to make this language and its logic explicit, always clearly representing in its terms, the need to intervene against, and redress, unjustified interference with life forms. With what language-games does one disapprove of, or reject, unjustified human interference with nature?

The evaluative terms "bad" or "defective"—as in "the finless shark is defective" or "flailing around while drowning is a natural badness of the form"— are too 'thin,' too abstract, and general. We need eventually to derive thicker, more specific descriptions to guide and differentiate meaningful interventions. Eventually, talk of natural goodness or defect must become more specific, and expressed by thick terms that motivate and guide specific human action in real-life contexts. We have already used the word "cruel" in this connection. There are many others. Once the language-game is capable of being played at this level, it makes sense of other language-games that can serve as interventions too: e.g., the language-games of blaming, repudiating, ostracizing, and shaming. Recent literature in philosophy and the social sciences on shame is interesting and important here, showing its neglected power to influence us [6]. *Save the Whales* (Boxes 20.1 and 20.2) could be cited as a practice that effectively used shame and related methods: public demonstrations against whaling in the 1970s were intended to bring shame on those responsible, particularly whaling governments and their representatives assembled at the International Whaling Commission [7]. The strength of the practice of life form conservation is that, when it shames, rejects, or repudiates, it is based on a system of values that is derived from the satisfaction of primary natural needs—needs as objective as the natural historical descriptions of a life form. The values are not simply cultural, religious, subjective, aesthetic, or purely instrumental values, nor are they vaguely expressed values, like "intrinsic values." What is condemned as "cruel," or in any other thick terms, is ultimately condemned on the basis of the more general and abstract, primary standard of natural goodness.

Social rejection, shame, withdrawal of trust, and humiliation are appropriate and especially powerful when motivated and justified by resistance to evil, the unacceptable, uncompromisable, and what is not subject to cost-benefit analysis. We have seen their effective use in the struggle for civil rights and in many other areas of human ethical transformation, including occasionally in the effort to protect other life forms, as in certain animal rights movements. The difference is that, in the conservation of life forms, the language-game of shame, etc., is based first and foremost on assessments of natural badness. A great variety of interventions we find in other practices will therefore not be the primary tools of life form conservation, for example, economic sanctions or legal condemnation, or the effort to refine or accumulate quantitative scientific data as support for a practice. Life form conservation simplifies its interventions by founding them on the Aristotelian necessities we can all identify insofar as we can identify life, and what a living thing is, has, and does. Of course, our knowledge of these can be quite extensive, like that of the expert, or relatively limited like most of the rest of us, but they all serve immediately to condemn in a way that can reach everyone's conscience, insofar as they see themselves as one life form among so many others.

An Invitation

We invite all conservation practitioners to ask themselves, in light of the language of life forms, why they do conservation, and why they were moved to commit to it in the first place, if not for the sake of *natural goodness* (though one might not know it by that name). We ask this especially of those concerned with conservation of species, who must consider how any intervention based on values secondary to the natural goodness in the forms of life can have anything but an accidental relationship to curbing or stopping the threat of provoked extinction. This is the relationship that the language-games of sustainable development, or conservation of species as populations, have to that threat.

We have presented an alternative, whose language expresses primary goodness, according to the logical grammar that represents living beings in relation to the forms they bear. We have argued that this language is necessary for any human experience or comprehension of the living, and that the standard implicit in it is the only reliable, rational, and effective standard for protecting life forms from provoked extinction. The practitioner of conservation then has at her disposal a logical grammar that she already uses, and which is essential in two crucial respects: (i) it is necessary to the understanding of her own form and (ii) to making sense of her particular experiences of the other living things. In the language of forms of life, in the values derived from the form-bearer relationship, we all have a basis for judging the harm we do to life, and for the primary justification for intervening to redress it. If conservation can accept, as we have urged, the objective and logical basis for this familiar and accessible representation of life, its practices and reason for being are likewise objectively and logically sound, with no need for compromising with any other standard.

We all think and create with words and, if Heidegger is to be believed, we are therefore the guardians of language, the home of being. We must all be guardians, in particular, of the the rigor and universality of logical grammar that describes the language of life forms and, therefore, of earthly life itself insofar as that language contains the standard for conserving it in all its diversity. The alternative is the constant threat of the 'error of Wuhan.'

©Andrew Guevara

References

1. Heidegger, M. 1946. Letter on Humanism. In *Martin Heidegger Basic Writings*, ed. D.F. Krell, 217–265. San Francisco: Harper.
2. Wittgenstein, L. 2009. *Philosophical Investigations* (trans: Anscombe, G.E.M., Hacker, P.M.S. and Schulte, J. Revised fourth edition by Hacker P.M.S. and Schulte J.). West Sussex: Wiley-Blackwell.
3. Campagna, C., and D. Guevara. 2020. Evaluating Climate Change with the Language of the Forms of Life. In *Climate Change, Ethics and the Non-Human World*, ed. B.G. Henning and Z. Walsh. Routledge Research in the Anthropocene.

4. Foot, P. 1978. Chapter 2: The Problem of Abortion and the Doctrine of Double Effect. In *Virtues and Vices. And Other Essays in Moral Philosophy*. Oxford: Basil Blackwell.
5. Cathcart, T. 2013. *The Trolley Problem or Would You Throw the Fat Guy Off the Bridge*. Workman Publishing Company.
6. Jacquet, J. 2015. *Is Shame Necessary? New Uses for an Old Tool*. Pantheon Hardcover.
7. Campagna, C., and D. Guevara. 2022. "Save the Whales" for Their Natural Goodness. In *Marine Mammals: The Evolving Human Factor*, ed. G. Notarbartolo di Sciara and B. Würsig, 397–424. Cham: Springer. https://doi.org/10.1007/978-3-030-98100-6_13.

24 Contrasting Life Form Conservation with Alternatives

This chapter summarizes the essentials of some prevailing models of conservation practice in order to illustrate what a shift to life form conservation would entail. The general profile of most alternatives to life form-based conservation has already been sketched in the preceding chapters. The objective here is to integrate these generalities and to illustrate what position each practice takes in the face of provoked extinction and the threats that lead to it. The conservation practices that dominate today adopted the secondary values that guide them by first abandoning the primary standard by which they represented-evaluated life. Consequently, the return we are here advocating for, to a practice of conserving life forms, implies a return to a familiar language and logic. In this sense, the change from prevailing practices to life form conservation is not onerous, although it is in a way revolutionary, since it involves quitting and replacing dominant practices at the most fundamental level. The essential change comes with attention to language and its logic, and particularly the normative implications of language: the language-games of "good," "bad" "right," "wrong," etc., and the thick terms that fill these out in specific contexts and applications of conservation.

The Negotiator Model

Conservation practices following what we will call the "negotiator model" are based on the weighing of costs and benefits, measured in terms of values mostly secondary to life forms. For purposes of comparison, these practices can also be represented as proceeding in four stages: (i) forming an empirically confirmed hypothesis of a threat of provoked extinction, (ii) cost-benefit analysis, (iii) compromised solution, and (iv) intervention. Following the model, the practice proceeds by testing the hypothesis that the relevant activity (e.g. industrial whaling) affects target species populations. Once the threat is adequately supported by empirical data, various costs and benefits related to the threat are assessed: market demand for the product, effects on employment, legislative effects, effects on management strategies, or

C. Campagna, D. Guevara, *Speaking of Forms of Life*, Fascinating Life Sciences,
https://doi.org/10.1007/978-3-031-34534-0_24

effects on the biology of the population involved, among other things. On the basis of an analysis of the costs and benefits, a solution is proposed that mitigates costs by means of some compromised intervention. The usual goal is to not alienate competing interests, to be open to human 'needs,' and, if possible, or at least in the declared intentions, to take into account the viability of the species involved.

The language of science is important in this practice. Scientific data often drive the process and provide the substrate for negotiation. This is of particular importance to the many conservation scientists and eco-pragmatists (Chap. 18). The language-games of economics, statistics, and population dynamics are essential to the formulation of hypotheses and their confirmation, and the grounding of interventions instrumental to the desired balance of costs and benefits. The practices typically do not prioritize primary over secondary values in balancing costs and benefits; instead considerations based on both values are counted as equally relevant, and, often, secondary values prevail as human benefit is considered the most pressing value. Compromises take into account a diffusely defined human 'welfare,' in which our life form needs are not differentiated from conveniences, or tastes, or religious, political, cultural perspectives not necessarily related to those needs. The various language-games get mixed up. The trophy hunter's interest, for example, is set against the viability of populations and the need for resources for National Parks. Or deforestation for agriculture and livestock development is analyzed for the effect on biodiversity but also for the effect on poverty alleviation, indigenous peoples' traditions, market demands, regional development needs, the interests of sovereign states, private property rights, etc. Negotiation can take place without making explicit the value systems in conflict. In pluralistic approaches, all competing values and standards tend to be on an equal footing. It is then possible to reach a compromise that ends the conflict without having to see it as a conflict of values.

Cost-benefit analysis works best when applied to values that can take an economic form. For example, in fishing, the costs and benefits of an evasive boat maneuver, which could avoid some unintended catch, versus a more traditional maneuver, which ends up with non-intended catch, but saves time and fuel, could be weighed against each other. It is the same with the use of gadgets that could reduce the unintended catch of sharks, rays, birds, dolphins, turtles, etc., which add operational costs but may decrease mortality. Attempts to introduce the language of intrinsic value into the cost and benefit analysis result in a language-game of incommensurable values and compromises. The result is, for example, fishing activity in one area, for the sake of use value, and creation of a protected area in another, for the sake of 'intrinsic' value.

When the 'benefit' of reducing or eliminating the threat to the life form is weighed against a high economic cost, the trade-off may involve delay in any intervention to eliminate the threat, until more information can be gathered as to how to reach a 'better' compromise. This is often in conflict with the so-called precautionary principle: better avoid an action that may have an unknown, high risk of impact on target populations until there is better data. The principle is often avowed but it is rarely applied in these conflicts, or else confused with other principles having to do with least cost. Finally, the negotiator model is prepared to trade off even extinction as a possible undesirable but inevitable consequence of development (see quotes from the Brundtland Report in Chaps. 1, 2, 6, and 7).

The negotiator model is of course favorable to the standards of sustainable development, which values economic growth—a goal that threatens life forms, without necessarily redressing the threat to many of the life form needs of humanity. The standard implicit in the Aristotelian categoricals of our form, or any other, is not relevant to the model, except perhaps by accident. Instead, the satisfaction of individual desires, or of the interests of certain groups in society—both which might even be contrary to the natural human good—can be counted as benefits. This is masked, however, by not being made explicit, and thus never confronted in the process. If a conservation practice has as its primary reason to inform and support development, as is proposed by eco-pragmatists (Chap. 18, [1]), the cost-benefit model is appropriate. Extinction in this context implies a loss of opportunity for use.

The profound incompatibilities of the negotiator model with the practice of life-form conservation are obvious. Conservation on the basis of a combination of both models is impossible. The negotiator model might be useful if what we seek are compromises, across a confusion of conflicting language-games. But it is not the place to look for new principles that guide the relationship with other forms of life, or our own, for their own sake, i.e., for the sake of what they are, have, and do as living things of their kind. The model may not even be human-centered, inasmuch as it does not distinguish 'needs' that may be secondary for our form, from real needs of our form. The conservation of life forms, in contrast, allows for no compromise or weighing of costs and benefit when it comes to respect for a creature's autonomous satisfaction of its life form needs, which permits interference with those needs only if required by the needs of one's own form, and only for certain ways of satisfying those needs. There is no negotiation with those who promote the 'benefits' of trophy hunting to satisfy human tastes for 'sport,' not to mention the cruel, systematic, and indifferent way it is done. Nor is the clearcutting of a forest negotiated for the profits of a lumber company. There is no such negotiating table. It is banned by the logic and grammar of the conservation practice. Poverty alleviation is concerned with real human need, on the one hand, but it can, therefore, not be justified exclusively on values secondary to human life.

Conservation Models Based on Individuals

We have discussed certain conservation practices that target animals because of their similarities to the human life form (see Box 14.1). Animal welfare or animal rights movements are the main examples of practitioners grounded in the 'exceptionality' of some forms of life. The idea is that these human-like animals—e.g., in being 'self-aware' or at least conscious of pleasure and pain, and 'intelligent,' etc.—have *individual rights or other individual ethical claims on us* due to considerations of their 'welfare.' Obviously, the conservation target here is quite limited, given the narrow scope of life forms that meet the human-like criteria. Almost all animals, and six of the seven taxonomic kingdoms, do not contain sentient species of the type necessary to sustain the moral framework of this sort of practice. As we have discussed at length, the practice is based on a language-game that borrows from the language-game of ethical responsibilities between humans.

The tendency to project human qualities onto non-human living beings extends to the legal sphere. The case of the elephant Happy is a good example of how far this can go (see Box 14.1). The legal arguments are supported by moral ones, based on the notion of 'personhood.' In practice, this strategy is generally unsuccessful. However, it has not been abandoned; far from it, inasmuch as rights are now being claimed even on behalf of non-living things in nature, such as mountains, rivers, and the sea [e.g., 2]. We must be grateful to the extent that the animal welfare and animal rights movements have succeeded in opposing unacceptable human treatment of other animals, and they may have also contributed to avoiding extinctions, such as with whales and whaling in the 1970s. But the conceptual and thus fundamental limitations of their language-games are clear. The grammatical constraints and limits are brought out well when we consider language common to the justification for and motivation of this practice, e.g., "humans can use language, sponges cannot" or "gorillas can feel pleasure and pain, plants cannot," or "primates are capable of abstract reasoning, reptiles are not." This leads to a certain grammar of "lack," which is then thought to justify the application of certain ethical concepts and responsibilities or not, to the various types of animals. It allows us to 'rate' certain life forms as more capable than others, in terms of the favored human qualities that matter to ethical considerations between humans. Thus, reptiles, lacking in so many of the right qualities, will not be worthy of moral consideration, or hardly as much as primates. Forget single-celled animals or fungi![1]

It is without doubt a great wonder, sublime and beautiful, and something to learn from, when we see what the puffer fish can do, as illustrated on the cover of our book. But it is a mistake to try to understand and appreciate it in anything but the grammar and logic that describes its own form. Anthropomorphic projection can be useful in capturing our attention and wonder, especially as children, but not as a basis of proper understanding or evaluation. On the language of life forms, better or worse, lack or capacity, can be judged only in relation to a form of life, not across forms of life or any other point of reference [5]. As Wittgenstein observes,

> It is sometimes said: animals do not talk because they lack the mental abilities. And this means: "They do not think, and that is why they do not talk." But they simply do not talk [6, §25].

This is an important point about the language-game of "lack," well illustrated by the propositions of natural history. "The penguin does not fly" does not describe a lack when describing the life form, any more than "Humans do not fly." These do not express lack in any sense relevant to primary values. The focus of animal welfare

---

[1] This is why it captures popular imagination when there are signs of 'a mind' even at this level, as discussed, for example, in [3]; all of a sudden there is moral 'reason' to respect them after all, where supposedly there was none otherwise (see also [4]).

conservation is often directed at the good of an individual in terms other than the form it bears, or not necessarily related to that form. This is why it is incompatible with life-form conservation, as are all other practices that try to focus on the individual independently of the form it bears.

Population-Based Models

The language of the representation of life integrates the individual and its form into a logically interdependent unit, unlike the language-games that would focus on the individual and its 'welfare' or 'rights.' Likewise, the logical grammar of life-form language does not permit, as the language of population-based practices does, the evaluation of the act of, e.g., trophy hunting, which kills the individual creature to be relativized to population abundance (or more specific parameters directly affecting it, like reproductive age), as a free-standing value. In the conservation of life forms, each individual is important *because each individual is a representative of its form of life* (not because it may be 'sentient,' etc.). Here there is not much new to add from what has already been discussed at length about population-based conservation.

Models Based on Intrinsic Value

The concept of intrinsic value has been discussed in some detail in this work, particularly in Chap. 13, where we saw that the concept is deployed in contrast to practices that base everything on the instrumental value of nature and species, rather than with a direct concern, including direct concern to prevent provoked extinction. One might expect that in practices which admit the concept, intrinsic value would have first priority and not be compromised. However, even some negotiator models that admit considerations of intrinsic value as among the 'benefits' that come into play in the cost-benefit analysis allow instrumental value to outweigh intrinsic value in certain contexts, as the Brundtland Report clearly implies (see the citations on Chap. 1). As usual, attention to language is not considered part of dominant conservation practices or interventions, and so the incommensurability or inconsistency of the language-games of intrinsic value and use, or instrumental, value goes unrecognized and the practice becomes gravely separated from any coherent set of values or avowed aspirations. The confusion is entrenched before anyone notices it, if they ever do.

The notion of intrinsic value, we saw, has family resemblances with the concept of natural goodness, in the sense that Foot specifies in a passage previously quoted (Chap. 12 and 17) but worth here repeating, at least in part: "[Natural goodness] is intrinsic or 'autonomous' goodness in that it depends directly on the relation of an individual to the 'life form' of its species" [7, p. 26–27]. This clarifies the grammar of the term "intrinsic" in the phrase "intrinsic value" and gives sense to "autonomy" as an underlying fundamental value in any concept of intrinsic value recognized by life form language. A practice of conservation grounded in intrinsic value in this sense is based on the standard of natural goodness. The grammar underlying the language-game of the practice is then incompatible with the prioritization of secondary values, particularly those of the negotiator models. Nor does it integrate well with the menu of language-games of pluralism: e.g., it is not possible to put autonomy on a par with the secondary values given equal consideration in pluralism. The practice of the conservation of life forms does not rule out that living beings can be represented and evaluated with other language-games; what it rules out is that all of them are justified as equivalent, and that autonomous satisfaction of life form needs can be reconciled with use value at the same level of priority. The practice of conservation on the basis of intrinsic value must be understood in terms of the grammar that describes the relationship between form and bearer, the primary language-game.

Aesthetic Models

Aesthetic values, like the beauty and sublimity of living things and the natural environment, are a fundamental part of conservationist practices that have led to some effective interventions against degradation and disfigurement of nature. The aesthetic language-game in conservation is perhaps the most powerful current influence on us in preventing provoked extinctions. The model deserves extended treatment of its own, with great potential for linking aesthetic language-games to the moral language-game. The literature is classic and vast. Our focus here is limited, necessarily, to a few critical observations. Ironically, for all its power and appeal, aesthetic language does not dominate the language of conservation practices that assume themselves to be influenced by aesthetic values [8], an example of the power of the language-game of economics, or use, to displace alternative value—making them seem too romantic and unrealistic.

But for the environmental movement, aesthetic representation and appreciation of nature and its objects might have been lost to a traditional conception that was built around art as a human product, and its iconic representation in the landscapes that dominated Western art through the nineteenth century [9]. The paradigm was changed in part by the contribution of the influential American environmental movement, represented by naturalists, essayists, poets, and photographers. Among the giants are Henry David Thoreau, John Muir, and John Burroughs and, more recently, Aldo Leopold, Rachel Carlson, David Brower, Ansel Adams, and Gary Snyder, who also draw from a variety of other models, including non-Western philosophy and religion [8]. Nature provides an inexhaustible source of opportunities for aesthetic appreciation, because, among other reasons, of the inexhaustible capacity of the living to repeat and multiply beauty and sublimity in its objects, with minimal alternations.

Aesthetic appreciation, however, is intimate, subjective, subject to cultural differences, the historical moment, fashions, or personal tastes. It might be easy to agree about the beauty of the bird of paradise, but there are too many insects that will never inspire general protection on aesthetic grounds.

We have already cited Leopold's prescient and elegant observation regarding values as yet uncaptured by language (see header in Chap. 5). The language of life forms, as expounded in the essay before you, lays down a grammar and logic that allows us to move into the next stages, including ethical ones. One does not find aesthetic language in Aristotelian categoricals, because aesthetic language does not obey the same grammar. But the two are compatible, and the theory of natural goodness can build on aesthetic language in a kind of progression not unlike the one suggested by Leopold. A careful and complete account could itself be quite beautiful and sublime, showing how to play a new and powerful language-game, including terms that thicken the more abstract and theoretical terms of life-form language. We would maintain that finding beauty in natural goodness is part of virtue, i.e., part of the natural goodness of our will and character [10, 11]. There is an important bridge between the two, one that the practice of life-form conservation does not dismiss [12]. But the sufficient and primary basis for conservation lies in respect for the natural goodness of the other life forms, whether they please us, or fascinate us, or inspire us, or not.

Concluding Remarks

We have here summarized some models for the practice of conservation that could then be contrasted with the conservation of life forms. To quit the practices based on these models as primary and to instead make life-form conservation the primary model is to prioritize a fundamental, necessary language-game, common to human beings, and especially familiar and useful to many conservationists already. The model directs us to something we already know and begin to master at an early age, giving proper place to the necessities that derive from the representation of life as a category. A key point of the chapter might be summarized by emphasizing the grammar of "need" or "necessity." A main difficulty with the dominant conservation practices lies in their tendency to play language-games with "needs" that are really based on preferences, or on what we are used to, or political correctness, or the inability to imagine other ways of relating to the natural world because of entrenched and uncritical assumptions about it and our place in it. What the language of life forms changes, and requires of us, is first attention to the language and the realization that comes from simply being able to represent how in fact things are with us, including confused and unnecessary. The prioritization of the language of natural history, and the true needs of life forms, described by Aristotelian categoricals, should feel like a homecoming, a return to where in a sense we always thought we were or wanted to be: one life form among others, living and letting live, according to a familiar, clear, and rigorous standard, objective and universal for all life forms.

References

1. Marvier, M., P. Kareiva, M. Marvier, P. Kareiva, and R. Lalasz. 2012. Conservation in the anthropocene: Beyond solitude and fragility. *Breakthrough Journal*. Feb 1. https://thebreakthrough.org/journal/issue-2/conservation-in-the-anthropocene. Accessed March 11, 2023.
2. Kolbert, E. 2022. A lake in Florida suing to protect itself. *The New Yorker*, April 18.
3. Money, P.N. 2021. Hyphal and mycelial consciousness: the concept of the fungal mind. *Fungal Biology* 125 (4): 257–259.
4. Geddes, L. 2022. Mushrooms communicate with each other using up to 50 'words', scientist claims. *The Guardian*, April 6.
5. Baker, N.E. 2012. The Difficulty of Language: Wittgenstein on Animals and Humans. In *Language, Ethics and Animal Life: Wittgenstein and Beyond*, ed. N. Forsberg, M. Burley, and N. Hämäläinenet, 45–64. New York: Bloomsbury Publishing.
6. Wittgenstein, L. 2009. *Philosophy of Psychology - A Fragment* (trans: Anscombe, G.E.M., Hacker, P.M.S. and Schulte, J. Revised fourth edition by Hacker P.M.S. and Schulte J.). West Sussex: Wiley-Blackwell.
7. Foot, P. 2001. *Natural Goodness*. Oxford: Clarendon Press.
8. Campagna, C., and T. Fernández. 2007. A comparative analysis of vision and mission statements of international environmental organizations. *Environmental Values* 16: 369–398.
9. Carlson, A. 2000. *Aesthetics and the Environment. The Appreciation of Nature, Art and Architecture*. London: Routledge.
10. Cafaro, P. 2001. Thoreau, Leopold, and Carson: Toward an environmental virtue ethics. *Environmental Ethics* 22: 3–17.
11. ———. 2013. Rachel Carson's Environmental Ethics. In *Linking Ecology and Ethics for a Changing World. Ecology and Ethics I*, ed. R. Rozzi, S. Pickett, C. Palmer, J. Armesto, and J. Callicott. Dordrecht: Springer. https://doi.org/10.1007/978-94-007-7470-4_13.
12. Tyler, O. 2022. *Natural Goodness and the Affective Ground of Judgment*. Santa Cruz: Ph.D University of California.

Leaving Things as They Were 25

Adopt a system, accept its beliefs, and you are helping to reinforce resistance to change …
F. Herbert [1 , p. 349]

All that we have labored to accomplish in this essay might be said to come down to this: the practice of conservation ought to be based on the standard implicit in the language of the natural history of living things. This practice—the conservation of the forms of life, as we are calling it—is based on the teleological language that describes the 'instructions' which each living being needs to follow in order to live according to its form. In describing these necessities, it implicitly also expresses the standard for judging the primary goodness or badness of any individual creatures bearing the form. This naturalistic language is guided by the logic and grammar of the representation of life—fragmented, confused, ignored, or lost, by the dominant language-games of conservation.

Fundamental to the conservation of life forms is the guiding schema "S is, has, or does F:" given content by the swimming of the shark, the photosynthesis of the plant, and all autonomous creatures, undisturbed by industrial fishing or clearcutting or any other threat of human-induced extinction. The essence of every intervention is to protect, maintain, or restore this autonomous existence, by intervening against the human beings and human institutions responsible for the acts that unnecessarily harm the other life forms. All justifications and motivations for redressing provoked extinctions are to be found in the normative teleology of natural historical judgments, where we find the objective and universal standard for evaluation of all life forms, including the standard for judging ourselves in relation to how we affect the other life forms. The conservation of life forms proposes something relatively simple: recognition of that which was always in view and that extinction annihilates forever.

C. Campagna, D. Guevara, *Speaking of Forms of Life*, Fascinating Life Sciences,
https://doi.org/10.1007/978-3-031-34534-0_25

The Leopard

The mid-1950s novel by Giuseppe Tomasi di Lampedusa dates back to the Italian *Risorgimento*. Garibaldian troops enter feudal Sicily in the nineteenth century. Soon the independent kingdoms fall and Italy is unified under the House of Savoy. In Sicily, the aristocratic Don Fabrizio Corbera, Prince of Salina and amateur astronomer, witnesses the revolutionary change in history. The prince foresees that the social class he represents will soon be reduced to insignificance. Political authority and economic power are shifting to the burgeoning bourgeoisie, which he cannot, and does not wish to, acknowledge or comprehend. The new generation, on the other hand, embodied in the figure of his nephew and successor, Tancredi Falconeri, quickly adapts to the circumstances. It is Tancredi who reveals to Don Fabrizio the essence of the moment: everything needs to change so that nothing changes. Power, intrigue, ambition, hypocrisy, the need for recognition—the essence of the human being remains intact while the context changes.

It is the tragic reality of conservation as we know it that its efforts to change everything for the better seem to have left intact the powers, greed, ambition, etc., which threaten extinction and destruction of nature, including our own threatened self-extinction. The theory of natural goodness does not immediately resolve anything, but instead must be tested in application to the hard reality faced by any theory. The difference is that, as we have argued, the theory before you at least begins with the primary value, the primary standard for evaluating what is good or bad for living things, and completely abandons dominant current ways of representing and evaluating living things and their kinds, which are based mainly on secondary values.

Recall a remark of Wittgenstein's, cited earlier (Chap. 2):

> Philosophy must not interfere in any way with the actual use of language, so it can in the end only describe it. For it cannot justify it either. It leaves everything as it is [2 , p. 124].

We have paired this with Foot's echo of it:

> I have been asked the very pertinent question as to where all this leaves disputes about substantial moral questions . . . The proper reply is that in a way nothing is settled, but everything is left as it was [3 , p. 116].

Our interpretation of these remarks, as applied to our work in this essay, is to say that the practice of conservation of life forms now depends on our at least correctly describing how things are: in particular, the representation of life, in richer and richer detail and in terms of what creatures are, have, and do according to their form. Or again, as Thompson says along the same lines about the nature of life forms: "I think our question should not be: What is a life-form, a species, a *psuche*?, but: How is such a thing described?" [4 , p. 62]. There is a rigorous grammar that guides and constrains this description and, if we keep to it, also prevents being led astray by a conservation of 'populations,' 'priority species,' or whatnot, which may or may not have anything to do with what is necessary for creatures to live and flourish as the kinds of things they are.

Resistance to Change

As writer Frank Herbert, author of *Dune*, says in the lines at the beginning of this chapter, resistance to change can be a consequence of adopting a system of beliefs. In conservation as it now is, we are attached to entrenched views and language-games that support many misconceptions and confusions about what must be done if we are really to conserve species, in their great diversity. To adopt life form conservation is in a way simply a return to the way things always were in the proper representation of nature, and this is not onerous or unfamiliar change, except for the many unnecessary and misleading beliefs and values we have adopted, and that have contributed to the crisis of provoked extinction. At one time or another, most of us have adopted the system and beliefs of conservation biology, or of sustainable development, or of the intrinsic value of life and perhaps all of the above. And in adopting these we resist change, despite the worsening state of the conservation of life, and despite the fact that we play derivate language-games that could not be played but for the language-game that makes it possible to describe life in the first place and which implies the standard which, as we have proposed in this essay, could bridge to an ethics of our relationship to nature, in the virtue that lies in recognition of and respect for natural goodness, wherever we find it.

The Language of Life Forms and Science

The *a priori* representation of life—the template or schema we must apply to our experience in order to identify living things and their forms—is not inconsistent with empirical science, whether biology or any other. It is the necessary backdrop to it. Without it, there would be no subject matter for biology or any other life science. In this way, the concept of a life form is more fundamental; it is logically prior to any empirical investigation. However, we have everywhere emphasized that experience is necessary to the content, the 'filling in,' of the *a priori* schema; experience informs the differentiation necessary for distinguishing species and their specific natural goodness. What the scientific monograph does, through careful observation, is inform and enrich, indefinitely, the detail of the shape of the life it describes. It proceeds empirically: by observation, verification, correction, induction, etc., all the while organizing all the empirical data according to the logical grammar of a form of life, generating the specific categoricals of the monograph.

It is precisely because of such work and knowledge that many of us have become conservationists, whether our work is done at the scientific level of the expert in an experimental discipline, or at the level of the naturalist in the field, or weekend gardener, or beach lover, or explorer of overgrown vacant lots. Perhaps, we never knew explicitly anything of the grammatical interdependence of the science being done and 'doing' conservation, nor, therefore, of the standard of natural goodness implicit in the description of the form. We were nevertheless intuitively guided by a sense of this connection, one we all begin to grasp at a fairly early age. Today, the sciences most relevant to conservation are in profound need of close attention to the logical structure of this language of representation-evaluation of life forms. And from close attention, at the most expert and influential levels, there is great potential for eventually breaking through to the general public, especially in the education of children, where life cycle models and nature documentaries are familiar vehicles of instruction already. But so long as science persists in language-games that over-emphasizes certain matters, while ignoring key distinctions, like those between provoked extinctions and the extinctions due to natural causes, rather than describing things as they are, the crisis will continue and only deepen.

The conservation of forms of life depends on a language-game for developing virtue in a new direction. This is where the role for education is clearest, since the work of developing virtue must begin early on. Virtue requires long practice, but the recognition of it, as recounted in the preface for conservationists at the beginning of this work, may be triggered by one or a few key experiences. The viability of our proposals for conservation depends, in its most elemental aspect, on the ethics of recognizing and respecting natural goodness in all forms of life, including especially all the non-human forms we now routinely and unnecessarily threaten with annihilation, at unprecedented rates. It depends on recognizing that living with respect for the natural goodness of any form of life is at least the beginning of virtue, and that ignoring, disregarding, or dismissing that value for the sake of wet markets, or economic development more generally, and so on, is defect or vice [5]. Here we echo, and offer rigorous grounds for, ancient ethical and religious traditions (mostly non-Western, or Native Western) that embrace responsibility for the most familiar thing in the world, natural life in its great variety of forms.

The theory we have presented, developed, and defended will confront objections and hard cases, like any other theory, including age-old questions skeptical of virtue or objective ethical value and so on. Our task here is to at least get the attention of those who have some heartfelt, intuitive appreciation for our idea that the primary value for conservation lies in the natural goodness of living things in all the great variety of their forms of life, including our own. We have demonstrated a rigorous basis, in a familiar language, for joining us in the conservation of life forms and the development of the human virtues that can be the only lasting basis for eliminating the threat of provoked extinction, including our own extinction.

References

1. Herbert, F. 1981. *God Emperor of Dune*. https://todaynovels.com/god-emperor-of-dune-by-frank-herbert-pdf-fre-download/ Accessed March 11, 2023.
2. Wittgenstein, L. 2009. *Philosophy of Psychology - A Fragment* (trans: Anscombe, G.E.M., Hacker, P.M.S. and Schulte, J. Revised fourth edition by Hacker P.M.S. and Schulte J.). West Sussex: Wiley-Blackwell.

3. Foot, P. 2001. *Natural Goodness*. Oxford: Clarendon Press.
4. Thompson, M. 2008. *Life and Action: Elementary Structures of Practice and Practical Thought*. Cambridge: Harvard University Press.
5. Cafaro, P. 2001. Thoreau, Leopold, and Carson: Toward an environmental virtue ethics. *Environmental Ethics* 22: 3–17.

26 Objections and Misunderstandings

The most obvious way in which our essay has left everything as it already was becomes evident when the pressing issues of the day are considered. What does our theory imply about diet and consumption generally? Whom should we boycott? What about those insects in our homes that are not directly attacking us? Should we have smaller families?[1] Stop using air travel? What are the implications of our view for abortion? Is it OK to eradicate forms of life that cause horrific diseases, like malaria or leprosy, particularly now that forced climate change may aggravate the risk that infectious diseases represent for humans [1]?

We have also left unaddressed, and thus as they were, the more general, theoretical challenges to Foot's attempt to base morality on natural goodness, which she herself only began to address—challenges raised by a certain reading of Nietzsche on morality, for example [2, Chap. 7]. Moreover, Foot limited her discussion to the moral relationships between human beings, and was at best ambivalent about extending the scope of our ethical responsibility to the other life forms. In the work before you, we have tried to make a contribution that goes beyond the way Foot left things in this case, and have sketched a conception of morality—based on natural goodness—that could guide and motivate our relationship with the other life forms and serve as a powerful corrective to provoked extinction. So our own work is doubly vulnerable: first, in adopting Foot's view of morality as the natural goodness of human will and character and, secondly, going beyond Foot in offering the beginnings of an account of how morality, based on natural goodness, might guide

[1] It is worth noting that the theory of natural goodness does not imply that it is always good to have children, or to have as many as possible, or even necessary to have any at all (see the treatment of this topic by Foot [2, p. 42]. Not every natural good must (or can) be instantiated in a good human life. Plus, of course, it is necessary to take care of the children and to secure a good life for them too. The reason we are in the predicament we are in, with human population size, is due to lack of attention to natural goodness, human, and otherwise. What to do about it now is one of the hard ethical questions, among many others. We have only begun in this book to sketch the foundational reasons and guidance for beginning to answer them.

C. Campagna, D. Guevara, *Speaking of Forms of Life*, Fascinating Life Sciences,
https://doi.org/10.1007/978-3-031-34534-0_26

our relationship with the other life forms. On top of this, our view is also subject to the centuries-long and, nowadays, intellectually fashionable rejection of the possibility of an objective, naturalistic morality, in search of first principles.

We must let such questions be for now. They will take considerable wisdom and good judgment to answer—to the extent that they can be answered. They are not just intellectual puzzles that any of us could expect to solve just by thinking about and discussing them. They are questions that can only be resolved through development of the virtues and good judgment necessary to do so, virtues that prioritize natural goodness over other values, and that enjoy mastery of the language-game of the representation and evaluation of life. For now, we briefly address in this chapter a few issues that we think will help confirm and clarify key concepts, and avoid common confusions at the most fundamental levels of the theory that we have only sketched.

No One Would Ever Want to Be an Elephant Seal

An intuitive and common objection questions whether the satisfaction of life-form necessities, i.e., natural goodness, is really always good. The violent mating behavior of elephant seals has been cited as an illustration of the point [3]. This is a happy coincidence, considering that one of us spent decades studying the behavior of these seals and other marine mammals. It is part of the seal's form of life to engage in violent conflict during mating season, with the alpha male beating its rivals for females into bloody submission. The females themselves are sometimes subject to forced copulation. We have ourselves already cited 'horrors' from the insect world that make the violence of elephant seals pale in comparison (see Box 7.2). The objection is that some evidently harmful and bad things result from the violence, and the fighting is central to what these life forms do to satisfy the necessities of their form. On the theory, the satisfaction of those necessities is supposed to be natural goodness, but then it is questionable whether natural goodness is always truly good, let alone the primary good for the individual bearers of the form. In fact, the evidence seems to show that it is obviously bad for the 'victims,' e.g., the 'losers' in the social hierarchy of elephant seals: they are rejected, expelled from the reproductive context, to bear alone the healing of their wounds. And even the victors—e.g., the alpha male, etc.—suffer: fights leave them as bloody and exhausted as their rivals. We should emphasize that we are talking about what is involved in the *satisfaction* of the necessities of these forms of life, and not what happens from defective behavior. Natural goodness is supposed to be a benefit to the individual; Foot speaks of it as contributing to an individual's "flourishing," and not just as a characteristic way of life. However, what is happening to the individuals of these forms of life does not look like flourishing.

This objection has intuitive force in part because it is necessary to our own life form that we recoil from, and are repelled by, violence, bodily harm, killing, and so on, especially in life forms that seem to experience pain and suffering. There is something wrong with us if we are not, at least initially, disturbed by these behaviors. The objection is also grammatically effective inasmuch as it is undeniable that, in the language-game of natural goodness, the violent behaviors are often harmful to both the victim and perpetrator, inevitably involving *natural badness* for them both, in at least the form of bodily harm (whether or not there is emotional or psychological harm). For example, if a seal breaks a canine tooth during the fight, his ability to inflict damage on competitors may be diminished, or a wound may become infected, draining energy needed for reproduction.

The objection is skeptical of the idea that flourishing could be consistent with as much blood, sweat, and tears, as we see in certain forms of life, but then we only need to be reminded of the hard things that are inevitably attached to living out a *good human* life, one with, for example, love of family and friendship. But is it so obvious that anyone would want to be a human being rather than an elephant seal, given the hard things inevitably involved in the satisfaction of *our* life form needs? Imagine a self-sufficient alien species that did not require or form bonds of love and friendship like we do. They might rather take a regular beating as an elephant seal than expose themselves to the risk of heartache, rejection, and anxiety inevitably involved in learning to form and maintain the bonds of love and friendship—made worse by our self-conscious awareness of it all. Either way, human love and friendship are great goods and the thought experiment brings out the questionable assumption in the objection: namely that natural goodness is inconsistent with the hard side of life for other kinds, when we do not accept the assumption for our own kind of life.

Again, we are not talking about the suffering and harm that comes from vice and foolishness and other defects. The point is that one cannot live a good human life without love and friendship, and without being vulnerable to its hard side, even when no one does anything wrong or bad. The question of whether life forms, human or otherwise, could have been better than they are, or that some are better than others, is not a question about natural goodness, but about some other standard of goodness, supposedly better and imaginable, where the lion lies down with the lamb. This standard would itself have to be investigated in a more wide-ranging critical discussion.

In a way, our response is simply a reminder that every form of life inevitably involves its bearers in some harm and killing (or at least the regular risk of both), in the satisfaction of the necessities of the form (Box 7.2)—predator and prey relations being the most obvious example. Animals fight, even though it may well kill them. Peter Geach makes the point with a nice analogy to human virtue, when he observes that human beings need courage like bees need stingers, though each may get them killed [4].

One cannot therefore object to the theory of natural goodness by citing the hard side of the necessities of life for other creatures, if at the same time one must admit, as one must, that what a good human life inevitably endures rivals that of any the other life forms. Perhaps, someone will still wish to say that human love and friendship are necessities that are worth the downside, whereas it is unclear what makes it worthwhile in elephant seal life. But now we must ask what standard for comparing the life forms is in play here in making such a claim, or is it, rather, just a lack of understanding or imagination, or the usual human-centered prejudices in our conceptions of nature.

Evolution Undermines the Theory of Natural Goodness

Foot briefly notes that the concept of natural goodness ought not to be confused with evolutionary concepts, like that of adaptation [2, note 37]. Her position is that the notion of a function, in the context of what a creature is, has or does, can in general be represented without reference to the evolutionary processes that, among other things, result in the adaptation of the parts and functions of the creature, parts and functions that contribute to differential advantage in survival and reproduction. The objection to her position on this score ([3], and see also [5]) is that without knowledge of these evolutionary processes, we cannot really say why a creature is the way it is: why it has what it has or does what it does. But this is what natural goodness supposedly tells us. That is, by identifying the Aristotelian categoricals that describe the life form, we supposedly learn how and why a living representative functions as it does, and has what it has. As we have discussed throughout the essay, we learn that this part here—e.g., the webbing on a certain kind of squirrel—is for expanding the surface area to create resistance to air, gliding from tree to tree, or evading predators, among other functions. Categoricals also tell us that the heart is for pumping and circulating blood, and getting oxygen throughout the body, or that the bear is inactive in winter in order to conserve energy, etc. Evolution, on the other hand, is mostly concerned with explaining the creature's parts and activities in terms of how they contribute to the passing on of high fidelity copies of genetic information into the next generation.

And the objection can be put like this: if we say nothing about how these parts and activities evolve, when composing the monograph of the natural history of an organism, then how will our description of what it has and does find its biological meaning? How will we understand the living things, and kinds, independently of the evolutionary history that shaped them? And how can we determine their primary good? The objection continues by pointing out that natural selection blindly shapes the organs and activities that make some living things more fit than their rivals or congeners for survival, reproduction and, ultimately, transmission of their genes. This same process also explains why nature is filled with the innumerable kinds of creatures that there are, rather than others that went extinct. Genes are the centerpiece of the process. But, again, how could primary goodness be determined without reference to this evolutionary account?

According to the objection, then, a natural history of the sort proposed by Foot or Thompson is not ultimately or primarily the way of understanding a creature; whatever a natural historical monograph might capture is largely not guided by knowledge of the processes of evolution; thus, its teleology will fail to capture what is most relevant: the explanation of why it is as it is, has what it has, and does what it does. It is only those processes and adaptations that tell us what a living thing really is and how it is supposed to function. And a timescale is required for representing the changes involved, and not just a snapshot of their form at a given time.

We might here distinguish two objections: one is that nothing about living things—their functions, characteristics, etc.—makes sense except in light of their evolutionary history, while the second relates to the idea that what is ultimately 'good' in an individual according to evolution is the transfer of its genes to the next generation, and not the natural goodness of individual living things. Both are a consequence of a certain view of evolution. As William FitzPatrick puts it, in expressing this view of living things:

> . . . from a natural functional point of view, given that they are products of natural selection, they must be regarded ultimately as gene replicating systems [3 , p. 114].

Our response is to call attention to a misunderstanding of the place of genes in evolution, which mistakenly *represents* individual organisms as though they were vessels for transmitting genes.[2] The mistaken picture is of natural selection as a process of 'selecting for' genes best suited for successful transmission (or 'selecting against' those that are not), when in fact the target of natural selection, what natural selection works on, is the *individual organism.* The theory of evolution by natural selection cannot be comprehended if the individual, living organism is bypassed. The individual organism is the object of selection [10], and the organism, as we have argued at length, must be understood in terms of the functions described by Aristotelian categoricals.

The metaphorical language of the "selfish gene" and "selection" can mislead insofar as it might seem to determine 'ends' for organisms, *or their genes*, in a sense that could compete with the ends described by Aristotelian categoricals. In fact, the teleology of categoricals and evolutionary propositions is not the same. Most important: natural selection is not teleological, as Ernst Mayr makes clear in an important paper on "teleology" and its various senses in biology and beyond [11]. "Selection" and "selfishness" are purposive terms in ordinary language, but in evolutionary theory, purposiveness in natural selection can only be a metaphor. As we have noted in Chap. 6, Robert Trivers explains why the metaphorical language-game of "selecting for selfish genes" was necessary, in communicating and developing a highly mathematical discipline, but such metaphors are only useful up to a point. If they give us the impression that the individual is dispensable, then the picture is not helpful. In sum, natural selection works on the *individual living thing;* it is the individual that is, at once, the target of natural historical judgments *and* natural selection. There is no incompatibility between the two accounts of life, natural goodness and natural selection. The language of each is grounded on the relation of form to bearer, and, thus, on the language of life forms, without which we cannot say or think anything about a living thing.

---

[2] This may be due to a misinterpretation of Richard Dawkin's metaphor of the "selfish gene" [6], and a misunderstanding of the foundational theoretical work he popularized [7–9].

Natural selection 'works' on, or 'selects' for certain traits, functions, activities, etc. of an *individual organism.* And, again, these are realities that can only be identified and understood in light of the wider context of the shape of the organism's life and thus in terms of the teleological language of life forms. There is no subject matter for the theory of evolution except for what living things are, have, and do as understood in terms of the form-bearer relation we have emphasized throughout, whose logical grammar is given by Aristotelian categoricals. Genetic material—considered as such, in isolation from the form of life in which it exists, does not *do* or *have* anything that can convey the concept of life, nor anything for nature to "select" or not. From a *logical* perspective, then, genes are relevant to evolution only because they can be related to the language of the forms of life—i.e., the language used to identify and describe life by naturalists from Aristotle to Darwin, long before anything was known about genes. The language of representation of life is not subsumed by the molecular processes; it is the molecular processes that are subsumed under the language of life forms (see Chap. 6, pp. 72–3).[3]

There is then an interesting and important parallel in the relationship between gene and individual organism on the one hand and individual organism and life form on the other. Just as we cannot identify or understand the individual except in its necessary relationship to its form, we cannot identify or understand the gene's role in evolution except in necessary relationship to the individual organism that 'carries' the gene. Natural goodness as a basis for life form conservation keeps the primary thing always in focus, the intimate relationship between life form and bearer, which is to say life itself.

Do We All Have to Become Jains Now?

In the ancient Indian religion of Jainism, monks and nuns show their respect for other life forms by, among other things, wearing cloth masks and sweeping the ground in front of them, in an effort to avoid harming the tiny creatures that surround us in the air or lie in our paths. Wouldn't we also be required to go to such lengths or worse on the theory of natural goodness, as interpreted and developed in this essay?

Natural goodness theory locates primary goodness in a very general relationship between individual creature and its life form, for all life forms and their individual bearers. Natural goodness does not prioritize one life form over another. Thus, if during the pandemic we wore masks to prevent harming or killing each other as humans, it would seem that we owe the other life forms a similar respect. But this, in turn, seems to lead to obviously absurd consequences.

This is a misunderstanding about the application and consequences of the theory. The response to it begins with the reminder that the natural goodness of a human being, like that of any other kind of life, consists in satisfying the necessities of its

[3]For further arguments addressing objections originating in a supposed conflict between natural goodness and evolutionary theories, see [12–14].

own life form and *not* those of any other. Satisfying those needs will inevitably have effects on the other life forms, some bad (some good), as we have observed, now, several times. There is not necessarily natural badness in eating what we or others have killed (as even vegetarians and vegans must do, in general), even though the results may be bad for the eaten. An agent or action's natural badness, as discussed (e.g., Chaps. 12 and 21), lies in killing or harming *unnecessarily,* i.e., without having to do so in the fulfillment of the necessities of its own life form.

What is imperative is that we acquire the virtue and judgment to discern when our killing and harming is simply from convenience, habit, unreflective cultural attitudes, ignorance, or many other possible blindnesses and rationalizations. This is the primary ethical challenge for anyone learning how to apply the theory, ourselves included, inasmuch as we are all lacking in virtue and judgment to some extent. At this altitude, the theory has only very general principles or standards to begin to guide us through all that we would have to consider. For example, general principles would have to consider that most people have very few resources, or spare time and energy to divert from taking care of their genuine needs, especially in places where there are many creatures that can harm or kill them. Also, we have evolved to have natural tendencies to be repulsed and repelled by certain living things and environments. Many of these may work to keep us out of trouble, where we must act fast, and, thus, in ways knowledge or rational deliberation could never do. The place for deliberation and action, and the experience and knowledge we need to inform it, is in determining, wherever we can, whether our killing or harming is truly out of natural necessity or not.

As philosopher Rosalind Hursthouse [15] points out in her review of Paul Taylor's *Respect for Nature* [16], the requisite respect for nature is not virtue until after long training, reflection, and habit result in an enduring character that can reliably express the right attitude across the innumerable complexities we encounter in life. Even if we understand that some of the principles that Jains carry to an extreme could be good for the living world, we cannot prescribe them as a general rule. A similar point holds for all the questions with which we began this chapter. It will take virtue and good judgment to get them right. And that is an invitation for all of us to work together in deepening our understanding of, and commitment to, the language-game of natural goodness: its meaning and implications for us as one life form among all the others.

References

1. Mora, C., T. McKenzie, I.M. Gaw, J.M. Dean, H. von Hammerstein, Tabatha A. Knudson, R.O. Setter, C.Z. Smith, K.M. Webster, J.A. Patz, and E.C. Franklin. 2022. Over half of known human pathogenic diseases can be aggravated by climate change. *Nature Climate Change* 12: 869–875. https://doi.org/10.1038/s41558-022-01426-1.
2. Foot, P. 2001. *Natural Goodness*. Oxford: Clarendon Press.
3. FitzPatrick, W.J. 2000. *Teleology and the Norms of Nature*. New York: Routledge.
4. Geach, P.T. 1977. *The Virtues*. Cambridge: Cambridge University Press.
5. Millum, J. 2006. Natural goodness and natural evil. *Ratio* 19: 199–213.

6. Dawkins, R. 1976. *The Selfish Gene*. New York: Oxford University Press.
7. Hamilton, W.D. 1964. The genetical evolution of social behaviour. I. *Journal of Theoretical Biology* 7 (1): 1–16.
8. Trivers, R. 1985. *Social Evolution*. Menlo Park: Benjamin/Cummings.
9. Smith, J.M. 1979. Game theory and the evolution of behaviour. *Proceedings of the Royal Society of London B* 205: 475–488.
10. Mayr, E. 2002. *What Evolution Is*. Basic Books.
11. ———. 1992. The idea of teleology. *Journal of the History of Ideas* 53 (1): 117–135.
12. Lott, M. 2012. Have elephant seals refuted Aristotle? Nature, function, and moral goodness. *Journal of Moral Philosophy* 9 (3): 353–375. https://doi.org/10.1163/174552412X625727.
13. Hacker-Wright, J. 2009. What is natural about foot's ethical naturalism? *Ratio* 22: 308–321. https://doi.org/10.1111/j.1467-9329.2009.00434.x.
14. Hursthouse, R. 2012. Human nature and Aristotelian virtue ethics. *Royal Institute of Philosophy Supplement* 70: 169–188. https://doi.org/10.1017/S1358246112000094.
15. ———. 2007. Environmental Virtue Ethics. In *Environmental Ethics*, ed. R.L. Walker and P.J. Ivanhoe, 155–172. Oxford University Press.
16. Taylor, P. 2011. *Respect for Nature: A Theory of Environmental Ethics*. Princeton: Princeton University Press.

Epilogue

To the naturalists from Aristotle to the present, who have taught us so much about the living world.—Mayr, E.[1]

We have argued that the extinctions that human beings threaten and provoke—affecting so many of the great variety of life forms—reveal a collective defect in us, an inconsistency with our own form of life. In almost any imaginable circumstances, it is not a necessity of our form of life to extinguish any life form. On the contrary, we have argued that it is a human life-form need to recognize and respect the good of any individual satisfying the necessities of its form. This is not an idealization but a plain fact about the true needs of our form and, thus, of what it is to be a good human being. We have called the language used to express these needs "the language of life forms" and argued that it implicitly contains the primary standard for making evaluative judgments about any living thing, including judgments that teach us what is it to be a good human being.

Any path forward in conservation, any progress in our thought and action, begins with language and must proceed with particularly close attention to the language-games and logical grammar that reveal our guiding values. The ethical advances, especially the pathbreaking ones—as in the struggle for human rights or in the environmental movements that have compelled us to question the human-centeredness of our values—support this point; their effectiveness lies in large part in successfully introducing new language-games. This, we have seen, is not simply a matter of introducing new words or phrases, but an underlying logical grammar that makes it possible to articulate the standards that identify, and motivate the redress of, the losses, harms, or wrongs we had grasped only intuitively and darkly before. "Natural goodness" is not just a new expression, but Philippa Foot's name for the value of an individual living thing satisfying the needs of its form, represented by the systematic, logical grammar of what Michael Thompson has called "Aristotelian categoricals." Our capacity to use this grammar is what makes it possible for us to identify and describe living things in the first place and to identify and redress the threats we pose to them and their forms of life. A conservation based on the

[1]Dedication to [1].

C. Campagna, D. Guevara, *Speaking of Forms of Life*, Fascinating Life Sciences,
https://doi.org/10.1007/978-3-031-34534-0

standards implied by this grammar would change radically the nature of dominant conservation practice, and yet, in a sense, there would also be nothing new about it; it would just be to allow a familiar language-game to do its thing: the language-game of naturalists that we all begin to learn at an early age and whose masters have taught us so much, from Aristotle to Audubon.

Our message is that the threat of provoked extinction can be overcome by our recognition of natural goodness as the primary value and by cultivation of the virtues of character necessary for enduring and reliable conservation of that value. Our proposal is that in the representation of life, according to the language of life forms and natural goodness, we find the grounds and motive for a categorical change for the better, in the relationship between humans and the other life forms.

A recent article in *The New Yorker* [2] describes the case of a patient who, in a therapy session, speaks of the experience of creating and taking care of a garden as "... the only time I feel I am good." The therapist draws attention to the construction of the sentence and to the difference between "*feeling* good" and "feeling that one *is* good." The latter transcends the individual subjective feeling of the former. What is suggested by the story is that the care of living things—and not necessarily those most closely related to us—can make one feel that one is doing what is expected of a human being.

In our own view, the language of natural goodness is not complete until we can account for the feeling of goodness. Kant investigates a similar idea about morality, when discussing the feeling of respect for the moral law [3]. Such reflections on feeling would go beyond the scope of the present work, but they nevertheless indicate an important direction that the grammatical investigation we have begun here would go. We have maintained that being human means understanding and respecting the needs of other living beings, in all their forms. It is virtuous to have the desire to understand and respect them. This desire and its fulfillment bring one closer to what it is like to feel like a good human being: to feel that one is fulfilling the highest expectation that one can have of oneself and participating in the most ambitious and sublime aspiration for humanity.

References

1. Mayr, E. 2001. *What Evolution Is*. New York: Basic Books.
2. Mead, R. 2020. *Nature and Nurture*. The New Yorker. August 24.
3. Guevara, D. 2000. *Kant's Theory of Moral Motivation*. Boulder, CO: Westview Press.

C. Campagna, D. Guevara, *Speaking of Forms of Life*, Fascinating Life Sciences,
https://doi.org/10.1007/978-3-031-34534-0

Index

C. Campagna, D. Guevara, *Speaking of Forms of Life*, Fascinating Life Sciences,
https://doi.org/10.1007/978-3-031-34534-0